CLIMATE THEATRE

Climate Theatre examines how distressed environments are portrayed through theatrical and performance arts.

Despite variations in form and location, this volume unites diverse geographical settings through their shared concern with the effects of rising temperatures on land, sea, sky and beyond. Throughout history, theatre and performance have reflected on and presented society's most pressing moral, ethical and sociopolitical challenges. Today, various performance traditions—from ritual practices to cultural activism and historical reenactments—are being reassessed for their environmental resilience and ecological consciousness. The collection also showcases innovative contemporary works that position climate as a dynamic and meaningful presence within drama, performance, activism and art. These contributions reflect the global nature of the climate challenge while carefully articulating the nuanced perspectives emerging from different geographical locations.

This volume will particularly appeal to performance scholars, environmental humanities researchers, climate activists, theatre practitioners and students seeking to understand the intersection of artistic expression and ecological crisis in our rapidly changing world.

Denise Varney is Professor of Theatre Studies at the University of Melbourne.

Peter Eckersall is Sidney E. Cohn Professor in Theatre at the Graduate Center, City University of New York (CUNY), NY.

Jen Parker-Starbuck is Professor of Theatre and Performance at Royal Holloway, University of London.

Routledge Studies in Theatre, Ecology, and Performance

Studies in Theatre, Ecology, and Performance (STEP) explores intersections between theatre, ecology, and performance studies in a global context. Amidst a volatile period in world history, the threat of environmental degradation—in particular, climate change, access to clean water, air, food insecurity, and species extinction—seems to hover just off stage. Yet theatre and performance studies are increasingly engaged with and add unique contributions to the debate about humans and the environment. *Climate Theatre* captures some of these new practices. It argues that the nexus between theatre and performance studies, and ecology, offer new ways of approaching the complexity of the climate crisis. This new synergy insists that climate art is never a singular felt experience.

Choreographing Dirt
Movement, Performance, and Ecology in the Anthropocene
Angenette Spalink

Theatres of Compost
Performance and Ecology for the Anthropocene
Abby Schroering

Climate Theatre
Ecology and Performance in a Climate Emergency
Edited by Denise Varney, Peter Eckersall, and Jen Parker-Starbuck

For more information on the series please visit the series page: https://www.routledge.com/Routledge-Studies-in-Theatre-Ecology-and-Performance/book-series/STEP#

CLIMATE THEATRE

Ecology and Performance in a Climate Emergency

Edited by Denise Varney, Peter Eckersall, and Jen Parker-Starbuck

Routledge
Taylor & Francis Group
LONDON AND NEW YORK

Cover Image: Climate Angels at Crib Point, Victoria, protesting a proposed gas terminal, 21 February 2022. Photograph by Julian Meehan. Image courtesy of Julian Meehan.

First published 2026
by Routledge
4 Park Square, Milton Park, Abingdon, Oxon OX14 4RN

and by Routledge
605 Third Avenue, New York, NY 10158

Routledge is an imprint of the Taylor & Francis Group, an informa business

British Library Cataloguing-in-Publication Data
A catalogue record for this book is available from the British Library

ISBN: 9781032953458 (hbk)
ISBN: 9781032953472 (pbk)
ISBN: 9781003584445 (ebk)

DOI: 10.4324/9781003584445

CONTENTS

ACKNOWLEDGEMENTS

The editors acknowledge the Traditional Owners of the unceded land on which this volume was originally conceived. We recognise the unique place held by Aboriginal and Torres Strait Islander peoples as the original owners and custodians of the lands and waterways across the Australian continent, with histories of continuous connection dating back more than 60,000 years. We pay respect to elders' past, present and future as well as acknowledge the importance of Indigenous knowledge in the academy.

The editors would also like to acknowledge the Australian Research Council (ARC), whose Discovery Project Scheme provided generous funding in support of two international symposia that gave rise to this volume. We are grateful for further financial and in-kind support that was received from the Faculty of Arts, the School of Culture and Communication and the Macgeorge Bequest at the University of Melbourne and The Graduate Center, City University of New York (CUNY).

Alongside the editors of this volume, our research team included Emeritus Professor Peta Tait (La Trobe University) and Dr Lara Stevens (Charles Sturt University). We thank Dr Kyle Harvey (University of Melbourne), our Research Assistant and Administrator, for his resourceful digging for references, his experience in publishing as well as his administrative and financial capacities that kept the project on track.

The impetus for this collection were two symposia: *Climate Collaborations: Art, Science and Future Scenarios* held at the University of Melbourne in September 2022 and *Climate Theatre: Understanding the Relation between Ecology and Performance in a Climate Emergency* held at The Graduate Center, CUNY, in November 2023. Contributors to the symposia responded to our invitations and calls for papers with a diverse wealth of stimulating and challenging new material on theatre and performance. Together, as the range of topics and locations indicates, participants offered new perspectives on climate change in diverse locations and in a variety of modes. The authors thank all the participants at the symposia.

In compiling this volume, we thank all copyright holders for their generous permission to reproduce photographic works within this book: Julian Meehan; Dear Climate; Peter Morin; Fast Familiar; Monira Al Qadiri; the 21st Century Museum of Contemporary Art in Kanazawa, Japan; Jill Orr; Matthew Newton; CLIMARTE; Maria Litvan; Dan Weiss; Anitra Nelson; and the Adbusters Media Foundation. We owe our thanks to the generosity of our colleagues Peta Tait and Lara Stevens for their support and expertise with various aspects of this collection; to Sarah Ann Standing for her enthusiasm for this collection; and to Laura Hussey and Swatti Hindwan at Routledge for their invaluable assistance.

FIGURES

CONTRIBUTORS

Katia Arfara is Assistant Professor of Theatre, Performance Studies at New York University Abu Dhabi and an independent curator. Her scholarship and curatorship focus on migration, ecology, memory and the politics of representation. Her research appears in *Theatre Journal*, *TDR*, *Contemporary Theatre Review*, *Theatre Research International* and *Performance Research*. As the Theatre and Dance Artistic Director and a curator of the Onassis Stegi (2009–19), Katia has initiated and curated numerous interdisciplinary festivals such as the Fast Forward. She is the author of *Théâtralités contemporaines* and the co-editor of *Intermedial Performance and Politics in the Public Sphere*. Dr Arfara co-leads the Elements Pod in the NYUAD 'Anthropocene: Urbanism, the Environment and Sustainability' Research Kitchen. Her book *Curating the Commons: Socially Engaged Public Art* is forthcoming from the University of Michigan Press.

Felipe Cervera is Assistant Professor in Theatre and Performance Studies and Director of the Centre for Performance Studies at UCLA. His current research program focusses on the politics of performance and outer space. He serves as Deputy General Editor of *Performance Research* and as a convener of the international network Social Studies of Outer Space.

Una Chaudhuri is Collegiate Professor and Professor of English, Drama and Environmental Studies, Dean for the Humanities and Vice-Dean for Interdisciplinary Initiative at New York University. She is a pioneer in the fields of Eco-Theatre and Animal Studies. Her recent books include *The Stage Lives of Animals: Zooësis and Performance*, *Animal Acts: Performing Species Today* (co-edited with Holly Hughes) and *The Ecocide Project: Research Theatre and*

Climate Change (co-authored with Shonni Enelow). She participates in collaborative creative projects, including the art collective *Dear Climate* and the working group CLIMATE LENS.

Elin Diamond is Professor Emerita of English and Comparative Literature at Rutgers University. Author of *Unmaking Mimesis: Essays on Feminism and Theater* (1997) and *Pinter's Comic Play* (1985), she edited *Performance and Cultural Politics* (1996) and co-edited *Performance, Feminism, and Affect in Neoliberal Times* (2017), *The Cambridge Companion to Caryl Churchill* (2009) and *Robbie McCauley: The Struggle Continues* (forthcoming).

Minty Donald is Professor of Contemporary Performance Practice at the University of Glasgow. She is an artist-researcher who works with matter or entities often described in global minority cultures as 'nonhuman', most recently rivers and rocks. Minty's practice-research is an attempt to both face up to and unsettle engrained human exceptionalism and its implication in extractive practices and the climate emergency. Minty regularly works with a longstanding (human) collaborator, Nick Millar. Recent public works include *Piel de la Ciudad/Skin of the City* (2024–ongoing), *With These Hands...* (2021), *Erratic Drift* (2019–ongoing) and *Aguas Ocultas, Aguas Olvidadas* (2018–21).

Peter Eckersall is Sidney E. Cohn Professor in Theatre at the Graduate Center, City University of New York (CUNY). His recent publications include *Dramaturgy to Make Visible: The Legacies of New Dramaturgy for Politics and Performance in Our Times* (2024), *New Media Dramaturgy* (co-authored with Helena Grehan and Ed Scheer, 2017) and *Performativity and Event in 1960s Japan* (2013). Peter is the resident dramaturg of the Not Yet It's Difficult performance group (Melbourne).

Milo Harries recently concluded his doctoral research at the University of Cambridge. Milo is currently a postdoctoral researcher at the University of Manchester, studying climate resilience narratives in the catchment of the River Aire. He is also an opera singer and coach and has worked at the Royal Opera House, Glyndebourne and Opera North.

Deborah Hart is a climate and environment-focussed creative producer and writer based in Narrm (Melbourne) and a co-founder and the current Chair of CLIMARTE. After 16 years working in development roles with leading Australian arts and culture organisations, Deborah left her profession in order to devote more time to climate activism. Deborah founded LIVE (Locals Into Victoria's Environment, 2006) and later co-founded CLIMARTE (2010) and ClimActs (2013). She is the author of *Guarding Eden: Champions of Climate Action* (2015).

For her role in sustaining a vibrant creative community, in 2023, Deborah was honoured with the City of Yarra's 'Contributions to the Arts' award.

Linda Hassall is Co-Director of the Performance and Ecologies Research Lab (P+ERL) and Deputy Program Advisor for the Bachelor of Arts Humanities at Griffith University, Australia. Her research focusses on using theatre processes to explore climate change events and ecological recovery and in building sustainable production process for transitioning the theatre industry to sustainability. Her recent book, *Theatres of Dust: Climate Gothic Analysis of Contemporary Australian Drama and Performance Landscapes* (2021), was reviewed by Stephen Carleton as 'leading the field in contemporary analysis of Australian climate change theatre'.

Kyoko Iwaki is Associate Professor at the University of Antwerp, Belgium. She is recognised as a Performance Studies scholar specialising in Japanese contemporary theatre focussing on the topics of catastrophes, ends and ghosts. Kyoko has published in journals such as *New Theatre Quarterly*, *Studies in Theatre and Performance* and others. She is an associate editor of *Performance Research.*

Eero Laine is Associate Professor and Chair of the Theatre and Dance Department at the University at Buffalo, State University of New York. He is a co-editor of *Lateral*, the journal of the Cultural Studies Association, and a co-author of *Mourning the Ends: Collaborative Performance and Performance* (2025).

Carl Lavery is Professor of Theatre and Performance at the University of Glasgow. He has published numerous essays and books on theatre, performance and ecology, including *Rethinking the Theatre of the Absurd: Ecology, Environment and the Greening of the Modern Stage* (2015), *Performance and Ecology: What Can Theatre Do* (2017) and *An Idea of a Theatre Ecology* (2025). He is currently working on two new books, *Getting a Taste for Landscape Through Performance: Politics and Ecology* and *Constellations of the Anthropocene.*

Maria Litvan is a theatremaker and a doctoral candidate in Theatre and Performance at the Graduate Center, City University of New York (CUNY). Her research focusses on questions of embodiment, memory and transmission in the Américas. Maria's work has been published in *TDR: The Drama Review*, *Performance Research* and *BOMB Art Magazine.*

Cen Liu is a PhD candidate in Theatre and Performance at the Graduate Center, City University of New York (CUNY). Her research focusses on popular performance, material culture and Eurasia encounters, tracing how humans and objects become entangled and co-constituted. She has collaborated on digital archive projects with the CUNY Digital History Archive and the Museum of Chinese in America. Her research has been supported by the Folger Shakespeare Library, the Morgan Library & Museum and several Graduate Center initiatives, including the Public Scholarship Practice Space (PS2), Digital Initiatives and the Early Research Initiative.

Sarah Lucie is Assistant Teaching Professor of Theatre in the Humanities and Arts Department at Worcester Polytechnic Institute. Her research approaches contemporary performance and digital art through new materialism, ecocritical theory and posthumanism. Her current book project, *Acting Objects*, explores the critical eco-conscious potential of human-non-human relationships on the contemporary stage.

Anitra Nelson is a widely published activist-scholar in the Informal Urbanism Research Hub (InfUr-) at the University of Melbourne. An interdisciplinary social scientist, she also writes creative non-fiction and makes films. Her short documentary *Beyond Money: Yenomon* (2022) has won minor awards in international film festivals, https://anitranelson.info/creative-artist.

Jill Orr is a performance artist crossing between performances for live audiences and performances for the camera. She has produced iconic images reflecting an Australian perspective. Jill was represented in the inaugural Venice International Performance Art Week in 2012. She was awarded an Australia Council Fellowship to produce *Antipodean Epic* from 2015 to 2017. In 2020–21, Jill was represented in *Australia: Antipodean Stories* at Padiglione d'Art Contemporanea, Milan, curated by Eugenio Viola. *This Tree* was commissioned for Monash University Museum of Art's *Tree Story* (2021–22). Jill was represented in *Know My Name: Australian Women Artists 1900 to Now* (2021–22) at the National Gallery of Australia. Her *Mitre Lake* work was shown at Art Basel in 2022 and in the This Is No Fantasy gallery in Melbourne, where she is represented. Her work, *The Promised Land Refigured*, was shown at Linden New Art for Photo24 in Melbourne.

Jen Parker-Starbuck is Professor of Theatre and Performance at Royal Holloway, University of London. She is the author of *Cyborg Theatre* (2011), a co-author of *Performance and Media: Taxonomies for a Changing Field* (2015) and a co-editor of *Performing Animality: Animals in Performance Practices* (2015) and other book chapters/essays. She was the editor of *Theatre Journal* from 2015 to 2019 and a contributing editor to *PAJ* and the *International Journal of Performing Arts and Digital Media*. She is the Performance Lead on the Worldbuilding Team for the AHRC-funded grant CoStar National Lab and PI for the AHRC/BRAID-funded 'P3R: Performance, Participation, Provenance and Reward in Responsible AI'.

Christel Stalpaert is Senior Full Professor at the Department of Art History, Musicology and Theatre Studies at Ghent University, Belgium. She is Director of the research centre S:PAM (Studies in Performing Arts and Media) and a co-founder of the FWO-funded research network CoDa (Cultures of Dance). She was a distinguished visitor at the University of Alberta, Canada, in 2023 and published widely in the field, including in journals such as *Performance Research* and *The Drama Review*. She recently published *Performance and Posthumanism* (with van Baarle and Karreman, 2021) and co-edited the volume 'On Activation' with Eylül Fidan Akıncı for *Performance Research* (2023). She is currently writing a monograph on *Hyphenated Thinking: Performance (Studies) Activating Ecological Encounters*.

Sarah Ann Standing is Associate Professor in the Humanities Department at New York City College of Technology, City University of New York. She is Book Series Editor for *Studies in Theatre, Ecology, and Performance (STEP)* at Routledge. Sarah has published in volumes and journals including *Remembering R. Murray Schafer*; *The Routledge Companion to Theatre and Politics*; *Global Performance Studies* (*GPS*); *Readings in Performance and Ecology*; *PAJ: A Journal of Performance and Art*; *Theatre Topics*; *The Eugene O'Neill Review*; *American Theatre*; *Terra Nova*; and *Western European Stages*. She is also an active playwright and director.

Lara Stevens is Lecturer in English at Charles Sturt University, Australia. She is an author of *Performing Climates* (2025) with Eddie Paterson and *Anti-War Theatre after Brecht: Dialectical Aesthetics in the Twenty-First Century* (2016), an editor and translator (Fre-Eng) of essays by Hélène Cixous in *Politics, Ethics and Performance: Hélène Cixous and the Théâtre du Soleil* (2016) and a co-editor of *Feminist Ecologies: Changing Environments in the Anthropocene* (2018) with Peta Tait and Denise Varney. Lara sits on the executive committee of Plumwood Inc., the board that manages the rainforest property of the late Val Plumwood, as well as her literary and cultural legacy.

Peta Tait, La Trobe University, is an academic and playwright and a fellow of the Australian Academy of the Humanities. She has written 70 scholarly articles and chapters, and her recent books include: the authored *Forms of Emotion: Human to Nonhuman in Drama, Theatre and Performance* (Routledge, 2022) and *Theory for Theatre Studies: Emotion* (2021); the co-edited *Feminist Ecologies: Changing Environments in the Anthropocene* (2018); the edited *The Great European Stage*

Directors Volume One (2018); the authored *Fighting Nature* (2016); the co-edited *The Routledge Circus Studies Reader*; and the authored *Wild and Dangerous Performances: Animals, Emotions, Circus* (2012). She is a co-author of *Ecology and Climate in Theatre and Australian Performance* (2025).

Denise Varney is Professor of Theatre Studies at the University of Melbourne. Her research is primarily in modern and contemporary Australian theatre and performance with interests in ecocriticism, politics, modernism, feminism, history and the archive. Her latest monograph is *Patrick White's Theatre* (2021). Other books include *Australian Theatre, Modernism and Patrick White: Governing Culture* (with Sandra D'Urso, 2018), *Theatre in the Asia Pacific: Regional Modernities in the Global Era* (co-authored 2013) and *Radical Visions: The Impact of the Sixties on Australian Drama* (2011).

Kris Verdonck is a visual artist and theatremaker based in Belgium. His creations are positioned in the transit zone between visual arts and theatre, between installation and performance and between dance and architecture. His work is internationally presented in theatres, museums and festivals. Recent works include *IN VOID* (2016), an installation circuit about human absence, and *CONVERSATIONS (at the end of the world)*, based on the work of Russian writer Daniil Kharms. Verdonck created a choreographic performance in collaboration with ICK Amsterdam in 2019, *SOMETHING (out of nothing)*, dealing with humanity's spectral existence in an increasingly catastrophic environment. Verdonck's *PREY*, a music theatre production, opened during the 2023 Klarafestival in Brussels.

Clara Wilch is a postdoctoral fellow in the Environmental Humanities at Vanderbilt University and an associate of the Mahindra Humanities Center at Harvard University. Her interdisciplinary research concerns artistic and philosophical approaches to environmental studies and climate justice, with a focus on the circumpolar Arctic and performance.

Lisa Woynarski is Associate Professor in Theatre at the University of Reading. Her work connects performance and ecology from an intersectional lens, foregrounding decolonisation. She is the author of *Ecodramaturgies: Theatre, Performance and Climate Change* (2020) and the forthcoming *Performing Urban Ecologies*.

Introduction

1

CLIMATE THEATRE

Ecology and Performance in a Climate Emergency

Denise Varney, Peter Eckersall and Jen Parker-Starbuck

Theatre and performance have historically reflected the great moral, ethical, social and political issues of the past and present. Globally, many performance genres, be they connected to ritual practices, cultural activism and/or histories of past societies, are now reappraised for their sense of resilience and connections to caring for the environment. Taking as our subject the equally significant epoch of climate change, beginning with the Industrial Revolution and its ongoing effects on the global environment, *Climate Theatre: Ecology and Performance in a Climate Emergency* also recognises the appearance of a distressed environment in the literature and theatre from the turn of the twentieth century. While recognising strong variations in form, matter and place, this volume features diverse locations unified by global concerns with the effects of rising temperatures on the universal presence of land, sea, sky and galaxy.

This volume rejects the argument that sluggish or ineffectual responses to the climate emergency are the result of a cultural crisis or a lack of scientific or technological know-how. On the contrary, the contributors demonstrate that the politics of climate change is a constantly evolving struggle with humans at the centre. These politics are captured in a substantial contemporary repertoire that mediates the impacts of climate change, in modes that range from site-specific local to large-scale indoor/outdoor events, and from species endangerment to conservation. In a variety of locations, performances of redress covering decolonisation, conservation and disaster relief, register local instances as well as steps towards a less extractive, more ecological ways of imagining a more sustainable future.

The collected works here reflect the global reach of the climate problem, while articulating the nuances that emerge in different geographical locations

DOI: 10.4324/9781003584445-2

from coastal to inland areas. To this end, the collection features examinations of theatre and performance that capture these variations. The range of perspectives within this volume comes from Asia, the Pacific, the Americas, Europe and the Middle East, encompassing the developed North and the global South. There are oceans, mountains, wetlands and droughts, continents and islands—writings on diverse ecologies, each of which offers new thinking about ways of approaching, seeing, living and mitigating the here and now of climate—literally and symbolically. Bringing together distinguished scholars, early career researchers along with artists and activists, this volume offers a provocative collection of new research addressing the intersection of theatre and performance with the most profound environmental and ecological challenges of the twenty-first century.

The origins of the collection lie in two related symposia: the first, *Climate Collaborations: Art, Science, and Future Scenarios* at the University of Melbourne in 2022 and, the second, *Climate Theatre: Understanding the Relation between Ecology and Performance in a Climate Emergency* at the City University of New York (CUNY) in 2023. One of the key focus areas we attempted to cover was how the national relates to the international in the context of global warming. Partner Investigators, Professors Jen Parker-Starbuck (Royal Holloway, London) and Peter Eckersall (The Graduate Center, CUNY), delivered keynote addresses that brought the diversity of the United Kingdom, the United States and Japan to the project.

The lead question for both symposia was how contemporary performance engages with climate change, its manifest appearances, structure and form, its incipient secrets and extremes. These topics include adverse impacts such as rising temperatures, melting ice, oceanic warming, catastrophic floods and fire as well as the loss of animal habitats and the threat of extinction. The responses from the participants capture the way theatre and performance imagine and critically engage with a changing environment, finding new techniques of expression, remaking traditional form in the present, and creating activist performance in public spaces.

Many of the chapters included here propose new vocabularies of description and conceptual observations of practice, driven by performance-based writing about ecological theatre and performance. In addition to its analyses of theatre and performance, this volume also interrogates new ways of thinking about performance as an intervention in climate politics. One of the advantages that theatre and performance have over other art forms is its enactment in real time; an immediate aspect that contributes to its capacity to condense overwhelming events, such as environmental catastrophe, into salient images. The effects of climate devastation can be overwhelming, creating a desensitisation that makes it unable to imagine sustainable futures. However, climate catastrophe is explored more immediately in theatre and can push back against the morass of such hopelessness. By engaging with the facts, in both epic and intimate ways, and by creating embodied experiences of climate ecology, theatre fosters empathy and critical resolve and creates understanding at a human level. More broadly, the collection explores how performance has the potential to mediate the impacts of ecological disruption and change. How does performance give expression to and navigate the range of emotions that accompany such change and loss? What performative techniques are favoured by activist performances in public spaces that critique a lack of leadership for climate action?

As a collection, the commentaries and scholarly work that follow suggest that ecology and ecocriticism are now embedded in a wide variety of dramaturgies and scholarly works. Acknowledging the history of climate consciousness across the field, the CUNY symposium included a special live event—a conversation between Bonnie Marranca and Jen Parker-Starbuck that traced Marranca's ecological thinking. Alongside Una Chaudhuri's writings and university actions (included in this volume), Marranca and Parker-Starbuck's conversation reflected on the history of art and performance as it was intertwined with environmental concerns. Since the mid-1980s, Marranca has written about landscape, ecology, climate and catastrophe, and her essays in *Theatrewritings*, *Ecologies of Theater*, and *Performance Histories* (among others) have illuminated the ecological work of artists such as Gertrude Stein, Robert Wilson, Rachel Rosenthal, Meredith Monk and other artists who have performed with and about nature. Marranca's own life in the Hudson Valley has contributed to her thoughts on landscape and gardening, theatre and performance. Her words remind us of the underlying important questions that this book takes up:

> Reading the narratives of cultures and landscapes, I began to ask whether there wasn't a more worldly way of experiencing theatre. How do geography and climate influence a work? What are the ways in which plant and animal life, animate and inanimate entities, the natural and the artificial interact? How do biology and the body determine the human drama? Frequent travels abroad influenced this line of inquiry as I followed the expression of culture in different kinds of landscapes, the quality of air and light and water and wind; the design of doors, windows, parks, and textures of stone and wood. What effect do they have on the performing body, the perception of objects in space? And what of the diverse species of birds and plants and the spectacle of contemporary life in the new ecologies of cities? *Ecologies of Theater* is the beginning of my search for a 'theatrewriting' more conscious of the magnitude of performance worlds where landscape, myth, and cultural memory create and bear witness to all histories of life.
>
> *(Marranca 1996, xiv)*

In the chapters that follow, we interweave cultures and landscapes of human and nonhuman drama as humanity confronts the scale of environmental catastrophe.

At the same time, some of the contributions also deal with the minute, the barely seen and the unnoticed changes taking place that are nevertheless registered in theatre and performance.

About the Chapters

This volume is arranged around five themes and includes scholarly essays, conversations between scholars and artists and artists' presentations of and about their own work.

The first section of the volume, *Thinking about Climate and Performance*, begins with Una Chaudhuri, whose pioneering work on ecological theatre identifies a paradigm shift from human centred to a more nuanced situated environmental approach to the performing arts. Her keynote address began with a quote from Caryl Churchill's *Love and Information* that acknowledges the fear that accompanies the state of not knowing how to respond to the scale and unpredictability of climate change-related weather events. The question of how to prepare for the unknown, the unexpected and the likelihood of future distress gives rise, she argues, to a plethora of possible solutions, self-help manuals and the like that also contribute to anxiety. In response, Chaudhuri engages her students through irony in a performance entitled 'Dear Climate' designed to redress fear-inducing narratives with a three-step program. This fascinating program takes the concept of the more-than-human connection to the environment to a new 'anthropoScenic' familiarity and friendship.

In a related way, Minty Donald and Carl Lavery examine nine words common to theatre and performance discourse (for example, Practice, Acting and Performance) towards a more reflexive use of ecocritical discourse. Adopting a more self-critical approach to the world-making capacities of language, including scholarly discourse, 'Theatre and Climate Change: Constellating a New Lexicon' calls for reflexive use of language to suture the broken conjunction between nature and culture, and history and natural history.

Peta Tait further develops her work on understanding sensorial and emotional experiences of climate change, arguing for an embodied and sensory understanding of the human and nonhuman worlds. Her chapter discusses two works of Australian Indigenous theatre, *Namatjira: Written for the Namatjira Family (Aranda)* by Scott Rankin and Trevor Jamieson and *Dogged* by Andrea James. Tait's analysis shows how these works express the values of trust and hope arguing that these attributes need to be understood through a history of Indigenous culture that is bound up with species movement and Indigenous custodianship of Country.

The next section, *Provocations in the Anthropocene: More-Than-Human,* begins with Jen Parker-Starbuck's chapter 'The Sea Is on Fire: Machinic Crustaceans and Ecological Promises', originally presented at the Melbourne symposium as a keynote address. Reflecting on recent environmental crises in the Gulf of Mexico where in 2021 the sea was literally on fire, she considers the spectacle of 'Les Machines de L'île', a grand installation modelled on the nineteenth-century concept of the 'theme park', presented in Nantes, France. Focussing on patrons' and her own engagement with large mechanical animals and mythical sea creatures, this chapter reflects on Jules Verne's epic novel, *Twenty Thousand Leagues Under the Sea* (1870), to expand on the current environmental crisis and species extinction. Parker-Starbuck reflects on the power of the machinic or substitute creatures to both point to and, at the same time, override environmental concerns as well as the complexity of the human-technological-nonhuman relationship.

Continuing the discussion of climate and artworks, *Being Prey at the End of the World* is a conversation between the artist and performance maker Kris Verdonck and the scholar Lara Stevens. They discuss Verdonck's *PREY* that is based on the ecofeminist text 'Human Vulnerability and the Experience of Being Prey) by the philosopher Val Plumwood (1939–2008). In this text, Plumwood recounts her experience of being attacked by a crocodile while canoeing in the Kakadu National Park in Northern Australia. Stevens has undertaken detailed research in the Plumwood archive and was able to share her work with Verdonck in developing this project. *PREY* reimagined Plumwood's text as a three-part performance that was spoken, sung and, finally, danced by three generations of Belgian women performers and that drew on the dramaturgical structures of Japanese Noh theatre. Stevens and Verdonck's conversation revisits the research undertaken for the artwork and connects Plumwood's awareness of how being prey decentres the human tendency towards extractive domination over the environment. They explore how Plumwood's message is reimagined in Verdonck's performance.

Sarah Lucie's 'Ecospheric Affect in *Solo Pour Octopus*' features the remarkable spectre of a living octopus in an onstage aquarium. A live feed video of the octopus is projected behind the aquarium, accompanied by improvised music that responds to its movement. An audio recording of multiple texts offering facts about the life of the octopus as a species is included. Lucie's account of this immersive live performance addresses the challenge of how spectators responses might differ from human-acted theatrical work, especially around questions of identity. Ecospheric dramaturgy, on the other hand, offers new ways of perceiving actors and actants in ways that are more than just observations of the spectacle of the nonhuman for human consumption.

Felipe Cervera's 'Performing Lunar Climates' asks what we can learn from the incidence of the first plant to die on the moon in the context of Elon Musk's extraterrestrial ambitions. Drawing on a 2019 testing of the possibility of terrestrial life in space, Cervera turns to Jacques Offenbach's operatic *Le Voyage Dans la Lune* (that was first produced in Paris in 1875) and George Méliès's *A Trip to the Moon*, to consider the neocolonial aspect of space exploration. From these beginnings, the notion of an extraterrestrial aesthetics for the performing arts emerges, which if coupled with ecological significance, can contribute to discussions of an increasingly necessary celestial and space (post)humanities.

Finally, in this section, Cen Liu analyses Yesnia Thibault-Picazo's 'The Cabinet Anthropogenic Specimens' and 'Anthropocene Slam', the latter an event held at the University of Wisconsin-Madison. This chapter argues that contemporary versions of the cabinets of curiosity destabilise the epistemological frameworks of human supremacy over objects and things. The unsettling of human and non-human relations occurs as old and obsolete objects (over which humans can assert progress) are replaced by imagined futuristic objects that show the planetary impact of human activity including climate change.

The third section of this book focusses on *Environmental Violence and Inequity*. It begins with Elin Diamond, who is widely recognised for her scholarship on feminism and theatre, with her chapter entitled 'Slow Violence in Cherríe Moraga's *Heroes and Saints*'. Moraga is a *Xicano/a* playwright, who, as Diamond explains, bases her play on actual cases of pesticide poisoning, leading to severe birth defects, among women agricultural workers in California between 1978 and 1988. Applying an interlinked, feminist/ecological structure to assemble the interconnecting factors leading to the disaster, Diamond's perspective also take the reader to Rob Nixon's 'Slow Violence', the Brechtian *gestus* and to contemporary research that indicates unacceptable levels of pesticides among children in California's San Joaquin Valley.

Lisa Woynarski shows how many of the exigencies of the climate crisis are now largely determined by city life and notes the importance of a better understanding of urban ecologies. She introduces the work of Tahltan Nation artist Peter Morin. Her reading of his play, *Cultural Graffiti,* that took the form of a song cycle confronts us with the 'forced forgetting' of colonial pasts in city sites where the erased history of Indigenous people is recuperated through language and song. She then considers the participatory artwork *The Strategy Room* by Fast Familiar in which local urban communities can make decisions about how to support carbon reduction policies. Each work, she argues, asks us to imagine sustainable and equitable climate futures.

Katia Arfara, 'Mutating Performance: On Monira Al Qadiri's Speculative Petroaesthetics' considers the performances, sculptures, installations and films of Monira Al Qadiri, whose work is concerned with the irreversible damage to ecosystems of the oil-rich Gulf nations. Examining performances that could be perceived as a performative counter-monument to petroaesthetics, Arfara argues Al Qadiri's art speculates on a woman-dominated age after oil and patriarchy.

The next section, *Climate Risk, Activism and Eco-Performance*, highlights the analysis of artworks that posit a new ecocritical genre of performance. Here, we can see examples of how the locus of performance structurally, as a dramaturgical innovation, incorporates ecological awareness. Leading this section, Kyoko Iwaki's essay explores how environmental catastrophes are an existential historical presence in Japan. She argues for an ecological cosmology informed by Buddhist thought and showing an awareness of *myō*—the invisible and supernatural. As Iwaki shows, *myō* opens onto a sensitivity to the coexistence of human and nonhuman forms, and in the aftermath of the Fukushima disaster, the realm of *myō* has returned to the Japanese consciousness. Iwaki explores this through Okada Toshiki's installation-performance *Eraser Forest,* featured ecological and ghostly themes in an age of half-transparent life.

In 'Performing Across Time', Jill Orr, an Australian artist specialising in outdoor immersive performance, is in conversation with Peta Tait, a longtime advocate of the artist's work from 1979 to the present. In a way that is strikingly similar to Iwaki's chapter, Orr is similarly attentive to the ecological, long before it was named as such. Tait's questions draw out the subtle, molecular, geological and cosmic in her outdoor durational work that are often immersive in relation to the environment and essential to the choreography. What was once attributed to poststructuralism is now aligned with ecological thought. Orr's works have unswervingly explored the limits of the body in different locations that are often challenging.

Linda Hassall continues the focus on performance practice by proposing a dramaturgical framework for engaging with communities recovering from climate disasters. This chapter is focussed on research into flood-affected communities that are faced with the prospect of permanent change to their former ways of life. Based on Hassall's prior practice, especially in Queensland, Australia, the model sits at the interface of social science and applied theatre. Principles of community engagement (eco-connectivity) and eco-emotional wellbeing are among the ten principles of practice that she proposes.

Deborah Hart is one of the founders of the Climate Guardians, an ensemble with an ironic performative twist on the radical art of political protest. This chapter offers an account of key moments of protest in the ensemble's repertoire. Climate Guardians refers to their appearance as angels, dressed in long white robes with large, feathered wings that spread out from the upper body. This photo essay captures highlights of the Climate Guardians' performances over 25 years. One of their high points, for example, was their appearance at the United Nations 2015 climate change conference in Paris, COP21, attended by 150 heads of state and over 25,000 accredited delegates. The photos capture the power of appearance without words against the backdrop of their targets, from fossil fuel company headquarters to threatened coastal environments.

Eero Laine's 'Wilderness Acts: Dorothy Molter and Performing the Boundary Waters' examines 'wilderness' as a contested term in environmental politics and legislation. Focussing on the life of Dorothy Molter, who lived on the Isle of Pines, a small island that she owned in northern Minnesota on a small remote lake, Laine traces how the woman defied wilderness protection legislation and lived simply and whose simple life is remembered and reperformed today through theatre, a museum and books. This chapter queries the difference between Molter's refusal to leave her sustainable wilderness life and today's wilderness tours that treat the landscape as a performance they pay to see.

As this book shows, climate activism has created new research practices and a deep awareness of the importance of interdisciplinary thinking. We note that research in the field often links culture and community with an understanding that research is also a way of survival. Such research through the lens of performance takes numerous forms and is characterised by the way that it works between STEM, the humanities, the creative arts and cultural studies. Three essays in this section—*Culture, Community and Survival*—highlight these interdisciplinary approaches.

Christel Stalpaert's essay explores the educative function of performance in drawing together perspectives from the arts and sciences to better address the climate crisis. Introducing the collaborative art of Maria Lucia Cruz Correia, Stalpaert explores how working between artistic and scientific practice is a form of hyphenated thinking that can meaningfully reflect on climate-related challenges. Stalpaert describes her own participation as a hyphenated thinker working on the SZenne River project *STILL HERE.* Her essay develops a new materialist concept of art-science-activist worlding, while also reflecting on the position of arts scholarship and research in the Belgium academy. Her essay shows how such projects rethink research about climate change and the environment to include inclusive forms of knowledge and practice.

In '*Burning Man*'s Climate: Eco-Utopic Confluences and Contradictions', Sarah Ann Standing considers the history of *Burning Man*, the yearly neo-pagan festival in the Nevada desert in the western United States that increasingly embodies the pressures of global climate change in microform: a desert landscape with howling winds, extreme temperatures and water-shortage concerns. She makes a timely argument that while the festival emanates out of a white, masculinist, technology-infused zeitgeist, it offers a timely study of the contradictions in utopic eco-protest.

Finally, Maria Litvan's 'Tacuabé's Bow: Surviving through Absence' centres its critique of colonialism in the fate of a traditional musical instrument that mirrors the historic suppression and later revival of Uruguayan cultural identity. Embedded in the decolonisation of a nation, the Tacuabé's Bow's survival offers alternative ways of approaching impending extinctions.

In the final section of this book—*The Recalcitrant Politics of Climate Change*—we acknowledge that we need to be thinking and doing things differently, as we confront obstruction and wilful ignorance of climate change. The essays that follow are proactive in suggesting ways to move against this recalcitrance. To this end, Clara Wilch's essay 'Climate Control in terms of Gender,

Nature and Stories of the Snow Queen' connects the fairytale character of the Snow Queen with ways to confront the contemporary crisis of climate change. Exploring the feminisation of this archetypal figure, Wilch traces its appearance from Hans Christian Andersen's fairytale to Disney's popular animation franchise *Frozen*. The feminisation of ice and cold is explored in Wilch's analysis to suggest a new horizon of possibilities for facing the climate crisis.

Milo Harries considers the value of failure in a chapter on writer and theatremaker David Finnigan, whose works include *Kill Climate Deniers* and *Pilot Episode*. Under the provocative title, '"It doesn't work. Of course it doesn't work. But it doesn't matter that it doesn't work": David Finnigan's Dramaturgy of Errors', Harries traces the evolution of climate theatre from acting as a vehicle for communicating the state of the environment to one that empathises with feelings of helplessness in the face of the scale and complexity of climate mitigation.

In a similar mode of rejecting didactic and distressing scenes of environmental damage, Anitra Nelson argues for degrowth, a mode of advancing ecological sustainability and socio-economic inequity. On this model, the performing arts would take up attitudes of empathy and empowerment towards the climate crisis. Accordingly, her own works, *Beyond Money: Yenomon* and *Beyond Money: A Postcapitalist Strategy* are each comprised of narratives accompanied by photographic stills of landscapes and communities based on self-organising and collective sufficiency.

In closing, while we do not resile from the challenges facing the planet and how all living forms are impacted by climate change, we do hope the collection presents the potential of creative thinking and practice. Overall, our book gathers case studies and artworks that demonstrate the ways in which the arts resists extractive economies and necropolitics. The essays herein draw attention not only to the importance of activism and resolve but also to the power of creative thinking and the need for imaginative solutions to climate despair. *Climate Theatre* makes the case for theatre as a means, if not for effecting change, then for showing how performance, embedded in ecological awareness, offers creative perspectives.

Reference

Marranca, Bonnie. 1996. *Ecologies of Theater: Essays at the Century Turning*. Baltimore, MD: Johns Hopkins University Press.

Thinking about Climate and Performance

2

CAMPUS ANTHROPOSCENES

Ecospheric Knowledge Production as Ambulatory Performance

Una Chaudhuri

> I'm frightened.
> Just walk instead of driving and don't take so many hot baths.
> I'm frightened for the children.
> There were those emails those scientists, I don't remember the detail
> no, it didn't make any difference in the end
> no, I think you're right most scientists agree it's a catastrophe. The question is how bad a catastrophe.
> It's whether they drown or starve or get killed in the fights for water.
> I'd choose drowning.
> Are you really not going to take it seriously?
> I don't know how to.
> I don't know how to.
>
> Caryl Churchill, *Love and Information*, Act 6, 'Climate' (2012)

A notoriously challenging dimension of the climate change problem is the havoc it plays with our sense of scale, with the fact that we usually expect small actions to have small results, and big ones big results, but are now thrust into a reality where our small actions and habits have already, we are told, been having big, unexpected and terrible results, and also that we must hope that our small individual actions, conscientiously undertaken, *will* contribute to large *desirable* results. This crisis of scale installs an impossible topography between the personal and the political, between the local and the global, between individual action and collective consequence. That is the scalar derangement Churchill's characters manifest when they talk about taking fewer hot baths or driving less. Yet it's also immediately clear from what follows that they are aware that this

DOI: 10.4324/9781003584445-4

view is absurd, and—further—that scientific knowledge, even at the distance from which they receive it, via media scandals and reductive generalisations, assures them that individual actions cannot save future generations from the scientifically certified coming catastrophe. This crisis of scale leaves them with a crisis of thought: they don't know how to *think* about climate change.

The shaky realism of people like Churchill's characters, people like us, offers little help when climate change begins to be experienced not only as wicked problem but also an *emergency*, drawing more and more people into crisis conditions, through material catastrophes like wildfires and floods and drought and ice-shelf collapse and species extinction, but also epistemologically, as the natural world seems to be slipping beyond the reach of reliable knowledge systems, including—most alarmingly—those of the empirical sciences, once so reliable and confident. As a result, the standard political question—'What is to be done?'—is now increasingly shadowed by questions of how to manage the emotional turmoil—that mixture of anger, fear, shame, sorrow, guilt and despair—that surrounds this subject. Questions about how climate change is making us *feel*—the proliferating phenomena of climate affect—now form the basis of a whole new field—environmental psychology, with its own neologism, 'solastalgia', defined as a form of emotional or existential distress caused by environmental change, or as 'the lived experience of negatively perceived environmental change'.

At the same time, the question of how we *should* feel is rising to the top of climate topics, especially for artists and activists, for whom it is a question of ethical responsibility as much as one of emotional self-preservation. Donna Haraway's (2016) bracing injunction to 'stay with the trouble' has been welcomed as a robust response and strategy, rescuing many people from political paralysis by asserting the value of remaining alert and engaged in spite of the apparent intractability of this crisis. Almost a decade ago, the invaluable anthology entitled *Arts of Living on a Damaged Planet* launched the creation of what we might think of as a 'Toolkit for Staying with the Trouble': methods, skills and practices with which not only to live well—artfully!—on a damaged planet but also to do so in the service of a future of multi-species thriving (Tsing et al. 2017). In what follows, I describe a project that aspires to contribute to that toolkit, the art collective Dear Climate, and, in particular, a recent project by that collective entitled 'Signs, Wonders, Blunders'. (The Dear Climate collective was founded in 2013 by Una Chaudhuri, Fritz Ertl, Oliver Kelhammer, and Marina Zurkow. Chaudhuri and Zurkow are the currently active members.)

At its inception, Dear Climate sought to identify alternatives to the then-prevailing kind of climate messaging, which tended to be either sternly admonishing or sentimentally anthropocentric. While much climate art at the time was dedicated to raising consciousness and fighting apathy, we focussed instead on building empathy, fostering biophilia. We began by inventing a three-step program: Meet Climate Change, Befriend Climate Change and Become Climate

Change. More like steps in a dance rather than steps on a ladder, meeting, befriending and becoming Climate Change involved repeated and recursive acts of observation (meeting), fellowship (befriending) and identification (becoming) in relation to the more-than-human world of other animals, plants, landscapes, planetary forces. The steps provided a choreography for dis-anthropocentrism, swerving away from the habitual gestures of human exceptionalism. They seek new metaphors for the relationship between humans to more-than-human worlds, new relational models to replace the prevailing ones: the official, sentimental one—nature as mother—and the one in actual practice—nature as enemy. The model of friendship, less encrusted with dominant values and therefore less cliché-ridden, seemed capacious and pluralistic. There are many kinds of friendship—could we draw from them to build a larger and more nuanced repertoire of attitudes and alignments with which to begin rebuilding the complex and intimate relationships that have been so sadly diminished by modernity's systematic backgrounding of nature, treating it as mere scenery for the human drama.

The last of the steps—becoming climate change—has of course been unfolding well beyond our project, reflected not only in cascading catastrophes and extinctions but also in mass movements like Extinction Rebellion and Greta Thunberg's Fridays for the Future.

Closer to us, in the academy, the new field of Environmental Humanities is surging, while our own field of theatre and performance studies can now think *with* a growing repertoire of deeply searching works addressing our planetary realities.

For Dear Climate, the intensification and extension of climate chaos—the 'becoming climate change' of the culture and the academy—led us to intensify our idea of taking climate *personally*—this time by relating it to our work as educators. The project I describe below sought to reckon with the extraordinary culture-wide 'becoming' of climate change from the perspective of education and knowledge production.

'Signs, Wonders, Blunders' was a site-specific installation, designed in conversation with the physical and cultural identity of the space it would inhabit: an entire university campus.[1] The conversation it sparked was about how a program of creative dis-anthropocentrism and biophilia, like Dear Climate's, could inform academic experience beyond the classroom (Figure 2.1).

'Signs, Wonders, Blunders' took the form of wooden trail markers, or signposts, each with three arrows pointing in different directions. The signposts were located around campus; some were placed where they were quite conspicuous, at forks in the road or crossroads, others were more stealthy, 'hidden in plain sight' in pretty busy areas. Each signpost had three arrows with different words or phrases, and each one had on it three white bands with brief project information, and a question: the same question on each one (Figure 2.2). The question was 'What do I need to know for the planet to thrive?'

FIGURE 2.1 One of the 13 trail markers comprising Dear Climate's installation, 'Signs, Wonders, Blunders', at Appalachian State University. Photograph by Cheryl Zibisky. Image courtesy of Dear Climate.

This question was the work's first way of shaping the conversations it might spark. The question explicitly invoked the subject of **knowledge production**, using the nature of the site to locate the viewer firmly in the context of education. Second, the question **personalised** the subject of knowledge: the pronoun used being I—rather than you—the sign implicitly cast the spectator as a partner in inquiry rather than—as 'you' might have done—the passive recipient of settled knowledge. Depending on how often or how deeply a spectator confronted this question, it had the potential of generating *a personalised, individualised curriculum*, in which the particular needs and interests of the learner would carry as much weight as the task of learning. The question was an invitation to start assembling one's own intellectual toolbox for staying with the trouble.

The first-person pronoun also signalled the participatory and performative dimensions of the work: the signposts were intended as a scaffolding for

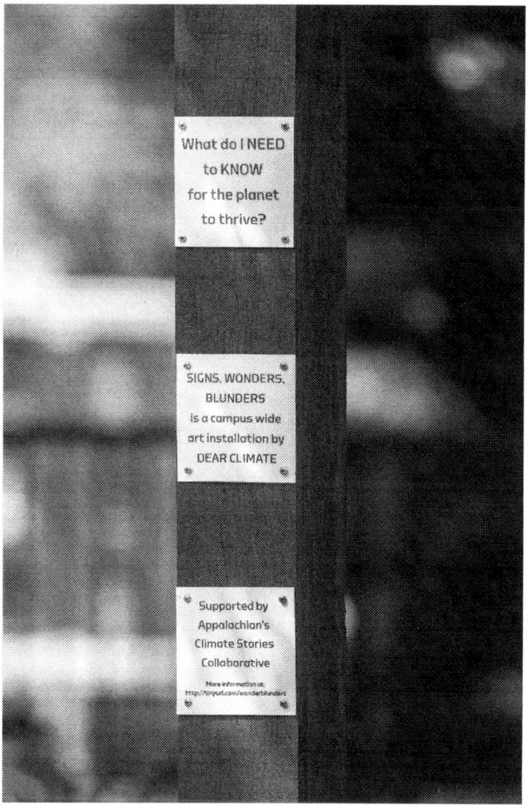

FIGURE 2.2 The question posed on each trail marker from Dear Climate's installation 'Signs, Wonders, Blunders' at Appalachian State University. Photograph by Cheryl Zibisky. Image courtesy of Dear Climate.

activities and activations, both pedagogical as well as creative, and ideally both at the same time. The installation would set the stage for a campus ecospheric performance, or what I've called, in the title of this article, 'Making the AnthropoScene on campus', inserting a theatricalising 'S' into that heavily contested and much-revised term. Elsewhere, I've described the AnthropoScene as *the drama of bad ideas*, staging—and thereby estranging—the *wrongness* of some strand of the thinking that has brought us to this catastrophe (Chaudhuri 2015). The AnthropoScene of 'Signs, Wonders, Blunders' was intended to confront that thinking—that is, the episteme of extractive modernity, including its hidden influence on intuitions like academia—with ideas and attitudes drawn from the burgeoning field of the environmental humanities.

The question also introduced—as a goal, as an ideal—the **thriving** of the planet. This goal contrasts with the goals more commonly mentioned in these contexts, such as 'saving' the planet, or pursuing sustainability, or adaptation, mitigation or harm reduction. The goal of thriving was in line with Dear Climate's Chthulucenic commitments and biophilic desires: the faith that there are emergent worlds beyond extractive modernity, and many possibilities for life in and beyond the ruins of capitalism.

Most importantly to us, the question implicitly defined the individual spectator's responsibility as a matter of *knowing,* rather than *doing*, a distinct departure from the emphasis found in most climate art and activist practice. To privilege doing over knowing is often to imply that what we already know what we should, and that the ways of knowing that produced that knowledge, are all fine, and sufficient, and appropriate, and that all we need to do now is take action, apply what we know. The Environmental Humanities of course, are deeply critical of dominant practices of knowledge production, along with the disciplinary hierarchies and institutional structures that those practices perpetuate. The university, once a bastion of settled knowledge with rigid divisions, is now itself in search of and in need of reinvention at the level of its divisions, hierarchies, methodologies and pedagogies. Experiential learning and flipped classrooms are examples of that ongoing reinvention, and 'Signs, Wonders, Blunders' sought to contribute to the same via new pathways and new directions suggested by current ecological thought.

Contemporary ecospheric discourse peppered the signposts in the form of titles of influential works (*How Forests Think*, *The Great Derangement*, *Staying with the Trouble*, *Laudato Si*, *Slow Violence*), sites of profound concern (the Larsen-B ice-shelf, the Great Barrier Reef) and key concepts (Buen Vivir, vibrant matter, alternative hedonism). Taken as a whole, these signposts constituted a kind of *dispersed curriculum* or *wandering syllabus*, cutting across traditional disciplinary divisions and including movement, especially wayfinding, as a key modality. One effect of linking ecospheric knowledge production with movement is to experientially shift the definition of environment from something that surrounds the human being, external to her, to something the human being moves and lives within. Tim Ingold (2009) has richly theorised this as a contrast between what he calls 'wayfaring'—the journey along a continuous line or along a pathway—and 'transport'—the movement from one self-contained point to another. The knowledge produced by wayfaring, says Ingold, is emergent, contextual, collaborative and embodied; that produced by transport is self-contained, portable and abstract. The knowledge produced by wayfaring is knowledge that grows, or is woven together with many threads; the knowledge produced by transport is assembled, made with building blocks. Now, signposts as such are usually tools of transport, pointing from one specific location to another; the signposts of 'Signs, Wonders, Blunders', however, encouraged wandering, doubling back or staying still. They hinted at the existence of hidden itineraries or new mappings, offering an embodied metaphor for the epistemic shifts and intellectual displacements that are needed if we want to align our current models of knowledge and education with the realities of our endangered planet. They invited people to keep going and notice **productive crossings**—between discipline, between species and across political boundaries.

Like the signposts, the title of the piece employed a three-part structure. Its first two words allude, of course, to the biblical phrase 'signs and wonders', but the third one, 'blunders', comes as a surprise. It interacts with the first two in ways that are also at work in each of the signs, setting up an oblique three-way dialogue which combines elements of cohesion, such as rhyme [trouble/rubble/bubble] or thematic repetition [forests-roots-branches, stories-fables-romances] with puzzling disruption [derangements-catastrophes-screw-ups]. Each tri-logue seeks to put its words and phrases into playful relationships that release other meanings, using the shiftiness of language, its capacity for suggesting meaning obliquely, through association, sound and tone. For example, while 'blunders' of course rhymes with 'wonders', it also introduces a different verbal register, more blunt, less poetic, and adds a note of scepticism to the idea of reading or following signs. It adds a subtext acknowledging all the mistakes and misunderstandings and sheer messy thinking that has led us to this moment of crisis and chaos (Figure 2.3).

FIGURE 2.3 One of the 13 trail markers comprising Dear Climate's installation, 'Signs, Wonders, Blunders', at Appalachian State University. Photograph by Cheryl Zibisky. Image courtesy of Dear Climate.

Wonder, the middle word of the work's title, the word that balances hopeful wayfinding against inevitable error, has a deep philosophical history that the Anthropocene will undoubtedly extend and revise. In a recent issue of *Environmental Philosophy* dedicated to responding to Dipesh Chakrabarty's (2021) influential theorisations, Urszula Lisowska takes up the idea with which Chakrabarty ends his text: namely, that this new planetary epoch that has erupted into human consciousness 'should lead us to revive the largely forgotten sense of awe and fear of the non-inhuman, in short: reverence for the planet' (2022, 270). Lisowska argues that the reverence Chakrabarty calls for is based on fear and results in a dangerous political inertia. In its place, she offers a politics of wonder: 'Being identified neither with mere curiosity nor with fear-based reverence', she writes, wonder should be understood as the attitude that creates familiarisation without appropriation. It is in this capacity, she says, 'that wonder becomes relevant to politics in the Anthropocene' (2022, 270). Also recently, Genevieve Lloyd concludes her comprehensive study of wonder from antiquity to the present, entitled *Reclaiming Wonder: After the Sublime*, as follows:

> My discussions of wonder in this book have highlighted one of its less obvious features: the ways in which it challenges ideals centered on certainty. Wonder, in its many forms, has provided an intellectual space—a temporary pause—which fosters the renewal of active, imaginative, emotionally engaged thinking. It depends but also thrives on the absence of certainty.
> *(2018, 214–15)*

The title of Lloyd's book echoes a passage from Hannah Arendt's *The Life of the Mind*: 'Reclaiming wonder can also mean something political: a concern with strategies of resistance to the spurious certainties which would have us ignore the crucial space between knowing and not knowing' (1971, 123).

Wonder and uncertainty were the modalities that we expected to rely on most during the activation phase of the work. During the semesters, the installation was scheduled to stay in place. Wonder as a noun (as in 'wonderment', a feeling or attitude), wonder as a verb, 'to wonder' in the sense of being curious, and finally, its homonymy with 'wander' were to intervene in the host location—the campus—and hint at alternatives to the goal of gaining mastery over a subject, alternatives like delightful private discovery, or thrilling shared insight. Its educational model favoured bewilderment over clarity, astonishment over understanding, playfulness over problem-solving and shaky grounds over certainties.

Conceived of as a **wayfaring practice of creative self-education**, 'Signs, Wonders, Blunders' positioned contemporary ecospheric discourse as an intellectual resource whose concepts and methods are marked by emergence and experimentation rather than by the painstakingly evolved systems and structures that characterise traditional disciplines. Thus, the project claimed unusual expansiveness and disciplinary flexibility for environmental thinking, for ecospheric discourse: a campus full of approaches to the more-than-human world and our relationships to it.

Quietly parodying traditional campus signposts, which typically assist in the traffic between disciplines—the science buildings, the arts and humanities buildings, the business school, the law school—these signposts created a citational map implicitly overlaid on institutionalised geographies of knowledge. For example, one situated near the science building obviously evokes a very different approach to matter than the one prevailing in departments of physics or geology. It leans on two influential texts of the environmental humanities, one of them foundational for the extraordinarily generative discourse of New Materialism. The third arrow, alluding to a television series but also to Jane Bennet's (2010) new materialist (and slightly spooky) concept 'thing power' offered both a challenge and an opportunity to look for points of convergence and divergence between different orders and scales of existence, from the vast to the ubiquitous to the uncanny, an opportunity to practice that most vital of ecospheric skills: connecting the dots. The three-part structure of each signpost suggested the possibility of a dialectical progression, in which opposites are resolved, but left its sequencing to the spectator (Figure 2.4). For example: can strangeness, or estrangement, be a way to deal with the derangements of scale resulting from

FIGURE 2.4 One of the 13 trail markers comprising Dear Climate's installation, 'Signs, Wonders, Blunders', at Appalachian State University. Photograph by Cheryl Zibisky. Image courtesy of Dear Climate.

climate change? Can a reenchanted world, where everything is alive, offer comfort and agency with which to face the vast unknowns of climate change?

The idea of a citational map laid over current academic mapping aligned closely with the agenda of one of the project's sponsors, a multi-disciplinary faculty project called the Climate Stories Collaborative. Like many initiatives nationwide, the Climate Stories Collaborative works to infuse ecospheric consciousness into all aspects of academic life and to encourage and facilitate interdisciplinary teaching and research among faculty. 'Signs, Wonders, Blunders' took the form it did largely because we knew it would be inserted into a campus-wide practice of co-learning and that many students would not only encounter it as they walked through campus but would also use it intentionally in classes and activities. Unfortunately, the COVID-19 pandemic caused 'Signs, Wonders, Blunders' to freeze in place. Those who would have used it, played with it, read the books it mentioned and made meaning from it: all went indoors and onto Zoom.

One obvious way to activate the work would be to deepen its site-specificity, making it an opportunity to ecologise that now-well-established genre of performance. The site-specificity already built into 'Signs, Wonders, Blunders' focussed on the institutional identity of the space, as a university. *Ecospheric* site-specificity would want to connect with many other aspects of this place: its flora and fauna, its geology and mineralogy, its demographics and their divisions, its histories and politics. Much could be done to connect this piece to an increasingly important cultivating local knowledge—material and cultural knowledge of a locality—both as a practice of biophilia and as valuable information for people and communities to use as we navigate the inevitable infrastructural disruptions and systems failures of climate chaos. In fields that have traditionally centred symbolic and figurative approaches to place and space, such as the arts and humanities, the move into materiality is as transformative and generative as was the earlier move into history, spawning new aesthetics, genres and practices.

Another inspiring form of ecospheric place consciousness relates to the 'Rights of Nature' movement, which draws from global Indigenous thought and environmental justice movements, to shift the legal paradigm connecting humans to the more-than-human world. In 2008, Ecuador became the first country to make the rights of nature ('Pachamama' or 'Mother Nature') part of the country's constitution. Prior to that, in New Zealand, the Te Uruwera forest, the Whanganui river and Mount Taranaki have been given legal status as living entities. Citing the Whanganui Act, a court in India granted legal personhood to the Ganges and Yamuna rivers. This also happened in Colombia, vis-a-vis the Atrato river. A site-specific assignment could use the movement as a model and invite students to select and honour some element of the campus's landscape, maybe even seek legal personhood for it (Figure 2.5).

Besides providing a legal way to protect the environment, these personifications dramatically expand the horizons of agency, interests and belonging beyond the human. They clear imaginal space for innovative relationality between and

FIGURE 2.5 One of the 13 trail markers comprising Dear Climate's installation, 'Signs, Wonders, Blunders', at Appalachian State University. Photograph by Cheryl Zibisky. Image courtesy of Dear Climate.

beyond species. We all agree that the two prevailing relational models—Nature as Mother and Nature as Enemy—are a bit shopworn. Dear Climate, as I mentioned, proposes the model of friendship; Bruno Latour (2014) proposed diplomacy, Beth Stephens and Annie Sprinkle (2021) brilliantly propose sexuality, as a model for approaching the more-than-human world in our time. 'Signs, Wonders, Blunders' proposes studentship. It attempts to link the goal of planetary stewardship to the practices of higher education in deliberately open-ended and experimental ways. If our encounter with the planetary confronts us with new tasks of wayfinding and world-building, it also presents us with an opportunity not only to mobilise all the fascinating complexity of studentship but also to expand and transform it. Along with desiring knowledge while fearing failure, along with the anxiety of influence and the elation of discovery, anthropoScenic studentship would welcome ambiguity and uncertainty, centre emergent knowledges without needing to master them and walk with wonder through this changing world.

Note

1 *Signs, Wonders, Blunders* was commissioned by the Smith Gallery of Appalachian State University, and developed there with the brilliant partnership of curator and faculty member Jenny Carlisle.

References

Arendt, Hannah. 1971. *The Life of the Mind: Volume One – Thinking.* New York: Harcourt Brace Jovanovich.

Bennett, Jane. 2010. *Vibrant Matter: A Political Ecology of Things.* Durham, NC: Duke University Press.

Chakrabarty, Dipesh. 2021. *The Climate of History in the Planetary Age.* Chicago, IL: University of Chicago Press.

Chaudhuri, Una. 2015. 'Anthropo-Scenes: Theater and Climate Change.' *Journal of Contemporary Drama in English* 3 (1), 12–27.

Churchill, Caryl. 2012. *Love and Information.* London: Nick Hern Books.

Haraway, Donna. 2016. *Staying with the Trouble: Making Kin in the Chthulucene.* Durham, NC: Duke University Press.

Ingold, Tim. 2009. 'Stories against Classification: Transport, Wayfaring and the Integration of Knowledge.' In Sandra Bamford and James Leach, eds, *Kinship and Beyond: The Genealogical Model Reconsidered.* New York: Berghahn, 193–213.

Latour, Bruno and Heather Davis. 2014. 'Diplomacy in the Face of Gaia: Bruno Latour in Conversation with Heather Davis.' In Heather Davis and Etienne Turpin, eds, *Art in the Anthropocene: Encounters among Aesthetics, Politics, Environments and Epistemologies.* London: Open Humanities Press, 43–56.

Lisowska, Urzula. 2022. 'Wonder and Politics in the Anthropocene: Beyond Curiosity and Reverence.' *Environmental Philosophy* 19 (2), 269–87.

Lloyd, Genevieve. 2018. *Reclaiming Wonder: After the Sublime.* Edinburgh: Edinburgh University Press.

Stephens, Beth and Annie Sprinkle. 2021. *Assuming the Ecosexual Position: The Earth as Lover.* Minneapolis: University of Minnesota Press.

Tsing, Anna Lowenhaupt, Nils Bubandt, Elaine Gan, and Heather Anne Swanson, eds. 2017. *Arts of Living on a Damaged Planet: Ghosts and Monsters of the Anthropocene.* Minneapolis: University of Minnesota Press.

Further Reading

Francis, Pope. 2015. *Laudato Si': On Care for Our Common Home.* Vatican City: Libreria Editrice Vaticana.

Ghosh, Amitav. 2016. *The Great Derangement: Climate Change and the Unthinkable.* Chicago, IL: University of Chicago Press.

Kohn, Eduardo. 2013. *How Forests Think: Toward an Anthropology Beyond the Human.* Berkeley: University of California Press.

Morton, Tim. 2013. *Hyperobjects: Philosophy and Ecology after the End of the World.* Minneapolis: University of Minnesota Press.

Nixon, Rob. 2011. *Slow Violence and the Environmentalism of the Poor.* Cambridge, MA: Harvard University Press.

Tsing, Anna Lowenhaupt. 2015. *The Mushroom at the End of the World: On the Possibility of Life in Capitalist Ruins.* Princeton, NJ: Princeton University Press.

3

THEATRE AND CLIMATE CHANGE

Constellating a New Lexicon

Minty Donald and Carl Lavery

Authors' Note

In order to mark our respective positions, Minty Donald writes in the first person and reflects on her practice with her longstanding collaborator, Nick Millar. Minty and Nick's practice explores their (human) interrelations with things and matter categorised as 'nonhuman', such as rivers and rocks. Their practice centres around an evolving suite of actions, or micro-performances, devised with nonhuman collaborators like pebbles or puddles. Each action is also described in and disseminated through a performance score. When discussing their practice, Minty uses 'us', 'our' and 'we' to refer to Minty and Nick, unless otherwise stated. Carl Lavery writes from the perspective of a lone scholar concerned with concepts and theories.

Argument

Almost 20 years ago, in *Theatre, Intimacy & Engagement: The Last Human Venue*, Alan Read said this:

> There is no more point in throwing ecology at theatre and hoping for the best, than there is in prematurely binding theatre and the political or social issues without addressing the serious problems of their more substantial relations.
> (2008, 250)

In recent years, there has been a sustained and rigorous attempt—sometimes directly, sometimes indirectly—to respond to Read's caution. As can be seen in

DOI: 10.4324/9781003584445-5

two recent special editions of *The Drama Review* in 2021 and 2023, this call for greater criticality and realism in environmental matters has led to a profound interrogation of different forms, genres and practices of theatre and performance in the West and beyond. While this work continues to be crucial, relatively little attention has been paid to the vocabularies that we, as critics and practitioners, have used to think, make and write about 'theatre ecology'–a term that, as we use it here, assumes that theatre is always already part of the world, earth and cosmos, and thus much more than a mere tool or vehicle for representing or showing climate change. The language used to conceptualise theatre is never abstract; it helps us to determine what theatre can do ecologically by establishing parameters, possibilities and limits.

Lexicon

A lexicon is a specific listing of words, often geared to indexing the disparate elements that come together to articulate a particular practice, medium or discipline. The words in a lexicon are like stars in a constellation at night. They are part of a general idea but stand apart from that idea, too, insisting on their singularity. They show any practice to be, on the one hand, integral, consistent and unified and yet, on the other hand, internally divided, heterogenous, opened to an outside that it can never detach from. Lexicons are neither static nor summative. Words fall out; words are added. New terms are needed, and old ones are re-rendered. The thing about a lexicon is that its apparent unity is never totalised; it is at best only ever a 'cut', a bracket, cosmos in chaos.

To create a lexicon for theatre in an era of climate change is to follow three divergent paths simultaneously. First, to insist on words and concepts that are inherent to, and already preponderant within, theatre practice and theory; second, to import technical vocabularies from the earth sciences; third, to repurpose discursive terms that have yet to be accorded theatrical value. The logic is to provide theatre practice with a new capaciousness and dexterity one that would allow it to respond to the complexity of the Anthropocene, a troubling and controversial signifier that is full of paradoxes, problems and obscured responsibilities, relating to 'greenwashing', capitalist imperialism and environmental racism.[1] One aspect of the Anthropocene that perhaps can be agreed upon is that it refers to an epoch where rigid binary oppositions between human and 'other-than-human' life forms have been undone. As such, if theatre purports to impact on what Bruno Latour terms the 'new climatic regime' of the Anthropocene (2018 [2017]), then it needs to enlarge its lexicon, to position itself differently, to suture the broken conjunction between nature and culture and history and natural history.

But how to re-invent that lost conjunction, especially when, as Donna Haraway is correct to insist, 'it matters what knots knot knots...what ties tie ties' (2016, 12).

A lexicon can 'knot knots' and 'tie ties' by alighting on those moments where extant theatre vocabularies point beyond their traditional anthropocentric meanings to an earthliness beyond them, while simultaneously, but inversely this time, reflecting on moments where words associated with the earth's systems reveal an in-built theatricality–stars, cosmos, actors, atmospheres, tipping points, catastrophes, environments, scenes, stages, plateaus, bios, etc. The point of this reciprocity is not to naturalise theatre and/or to aestheticise the earth in any essentialist sense, but to produce a reconfigured set of correspondences between them, to create encounters that would allow for a healthier ethos and habitus of planetary being. This would be one, moreover, that would represent humans as *climatic* as opposed to *climactic* creatures. That small, indiscernible shift in spelling is tantamount to a quantum leap in thought. It designates theatre and performance as exercises in planetary poesis, 'more than cultural' practices that dedramatise existence, calling into question our obsession with the actions of human characters alone. Theatre is a pluralist medium, an instrument that 'distributes the sensible' differently (Rancière 2004, 12). Not just because it widens the spectator's vision, as is commonly thought in most readings of that much-cited phrase from Jacques Rancière. But more pertinently, because it makes the spectator responsive, vulnerable and fragile, closer to things. This enhanced sensitivity, this strange intimacy with the earth, is what a new lexicon for theatre may be able to bring about. This is the type of 'relation' we are interested in knotting.

Stars/Stars

There is a close connection between stars and theatre, a double entendre of sorts. We can see this in the importance attached to such things as the 'star system', the operation whereby famous and charismatic performers are metaphorically associated with distant planets on account of how they bring radiance to our world, while remaining unapproachable. In the theatre, 'stars' have historically been objects of fascination and love; they seduce us, they make us curious, they catalyse desire. But today, that understanding of stars needs to be challenged. What if theatre were like a star because of its capacity to orientate and to direct, to produce a shape–a constellation–that we could follow, that might guide us, in the way that sailors use the pole stars? According to Walter Benjamin, the stars were the earliest form of theatre known to humans (Benjamin 1999, 685). The starry night was a *theatron* that performed a primordial act of mimesis, one that is still in operation but that we have perhaps forgotten, concerned, as we have been, since Plato and Aristotle, with an understanding of mimesis as an act of representation, the becoming equivalent of a thing. As Benjamin sees it, mimesis is not about depicting or establishing some visual *ratio*; it is about establishing a sense of intimacy with nonhuman life, 'a capacity for producing similarities' (1999, 721). Importantly, similarity does not mean oneness. A gap invariably remains. To think of theatre as a star is to return to that understanding of mimesis. To realise that performance can blur the boundaries between nature and culture while also keeping them apart. A star theatre is a theatre where the 'seen' co-exists with the unseen. In it, the Western episteme is troubled by an opacity, an enigma, that it is unable to turn its back on and wilfully erase from the picture. Just as the light that reaches us from distant stars is already dead, not quite what it seems, so the theatre troubles the metaphysical desire of the human to be on time, centred, a master of the universe. To wish on a star is to embrace an always theatricalised 'as if'.

Practice

Practice is the word Nick and I use to describe what we do as artists, and as researchers. I am conscious that practice, used promiscuously by artists and art students to describe a sometimes loose set of methods, a position and a context for their work, has come in for some criticism: for its ubiquity; for the arrogance of coopting a word once reserved for the disciplined, effort-full practice of medical doctors, lawyers, professional dancers and musicians; and for false dichotomies created between practice and other ways of working, such practice and research. But, for me, the word practice has valuable resonances and complexities. I want to keep practising. To practice is to work in a way that requires ongoing and repeated physical interaction with materials and places; with human and other-than-human matter and things. Practising is repetitive and attentive. Repeating with the aim of acquiring skills, body-memories and sensitivities that allow us to be responsive to unexpected turns and new situations. Practising trains us to notice small details and differences. To attune to fluctuations in our entanglements with other-than-human stuff. Practising requires improvisation, trying things out, making mistakes, learning through doing. We practise not to reach perfection, nor in pursuit of a polished presentation before an audience, but from a need to keep acting, to keep being attentive, to keep the faith. Practising is never completed. We continue to practise, in the face of climate change, mass extinction and environmental derangement.

Acting

In keeping with the Latinate etymology of the word drama, to act is to do. And in a world on fire, it comes as no surprise to find that we are constantly being asked to act, to take part, to do something that would combat the inequities and inequalities of 'slow violence' (Nixon 2011). But while there are moments and contexts when acting is absolutely necessary, there are other moments when it needs to be troubled. In his writings on Samuel Beckett, Steven Connor distinguishes between 'going slowly' and 'slow going'. Where 'going slowly' gives the impression that you are central, in charge of the situation, able to reassert agency by regulating the pace, ‘slow going’ complicates that agency, it positions you as a thing in a landscape of other things, a creature that is vulnerable, powerless, swept up and away (Connor 2014, 115–16). How to reconcile these two approaches to slowness, to learn that theatre is not just about acting but allowing oneself to be acted upon? Consideration needs to be given to this question. To think that we can act our way out of climate change is to perpetuate a fantasy of solutions that may perhaps only worsen the situation. By itself, slowing down could, in fact, speed things up. What if theatre's power were to reside in weakness, in desisting rather than resisting, in undoing as opposed to doing? How might that impact on drama? Who would be the agent of the act?

Performance

Nick and I describe our practice as more-than-human performance. By this, we mean that performance is something that humans and other-than-humans do, and do together in ways that are entangled, complex, asymmetric and always in flux. For example, a score for *Water Carry*, one of the first micro-performances that we devised, takes this form:

> Scoop water from a river or water source with bare, cupped hands. Carry the water as far from the river or water source as you can. Stop when there is no water left in your hands.

In *Water Carry*, the water in our cupped hands, the riverbanks or other terrain that we navigate and countless other nonhuman entities are considered by Nick and me to be our co-performers, who all have agency in influencing the form and duration of each micro-performance. The distribution of agency in our more-than-human performance practice is always, however, unequal, shifting and circumscribed by our intentions.

Performance is, of course, always already more-than-human. While Western theatre and performance traditions, with their broadly humanist underpinnings, have tended to focus on the words and actions of human protagonists, theatre and performance are, in Marlis Schweitzer and Joanne Zerdy's words, ‘littered with objects: props, costumes, sketches, prompt books [...] video recordings, architectural ruins, etc., etc’ (2014, 21). Schweitzer and Zerdy's scholarship seeks to challenge the anthropocentric bias of Western theatre, arguing that other-than-human theatrical elements should be considered as ‘active agents performing alongside rather than behind or in service to human performers’ (2014, 28). This, certainly, is not news to anyone who has trained and worked as a scenographer or theatre technician, or who has crafted props, costumes or scenery. Nor is it news to Indigenous peoples and other Global Majority cultures where binaries of animate/inanimate and human/nonhuman do not adhere.

Margaret Werry, for example, writes of the Māori practice of acknowledging the active role of the location where a performance will take place. She quotes Māori theatre practitioner, Ngamaru Raerino who says, 'The walls and the floors are the ears and eyes – you make sure they're happy with your performance' (Werry 2020, 206).

So, performance is always already more-than-human, and we humans are also always already more-than-human. In Jane Bennett's words, 'there was never a time when human agency was anything other than an interfolding network of humanity and nonhumanity' (2005, 463). Or, as Stacey Alaimo attests, 'the human body is never a rigidly enclosed, protected entity, but is vulnerable to the substances and flows of its environments' (2010, 28). Framing our practice as more-than-human-performance, we believe, compels us to attend to both the radical possibilities of recognising our more-than-humanity *and* to our enduring entrapment and implication in a human exceptionalist world view: the legacy of the humanist traditions in which we are immersed. Our more-than-human performance practice has the effect of turning us back to confront our human exceptionalism. In the performance score, *Rolling Stone Drift*, we make this invitation:

> Find a small stone in an urban area. Ask the stone if it will play with you. If it agrees, gently nudge or kick the stone along the street, observing where it goes and how it moves. Stop when the stone escapes.

The score and its enactment are intended as prompts to face up to the contradictions, complexities and power imbalances of being more-than-human. Nick and I ask a stone to play with us, while assuming that it will not understand. We may choose to take it, whether it 'agrees' or not, committing an act of coercion or, perhaps, a small-scale extraction.

Repetition

The idea of repetition is central to theatre's mode of operating. Not simply because theatre, to an extent, always repeats and so is unfinished, but also, and more paradoxically, because repetition comes before the actual performance. The anachrony of theatre explains, perhaps, why the French define a rehearsal as *une répétition*. These two opposed meanings of repetition are useful with respect to thinking of how theatre might engage with climate change. As everyone engaged with theatre knows, to repeat is not just to remain the same, to replicate the original. Rather, repetition transforms the original, opening it up to contingency and context. In this respect, repetition in the theatre would seem to embody an invisible force, something that is impossible to grasp as a discrete object of thought. To repeat is to gain access to the thermodynamic law of entropy, a state in which even when one seems to be doing nothing, something is happening. This happening takes place beyond the will; it can never be located, since it is in constant movement, the expression of a kinetic earth. To repeat, then, is to intervene in a cosmic and planetary process that one is unable to escape from, no matter what one does. Refraining from action does not keep things as they are; it is to actively worsen the situation, to accelerate it. Consequently, the theatre's inherent commitment to repetition, be that understood as an event that comes either before or after the performance, is a way of responding to climate change in a progressive manner. Why? Because theatrical repetition intuitively expresses and performs a process in which everything is in flux, constantly transforming, becoming other than itself. It knows that remaining the same is impossible just as it knows that any desire for wholeness is a futile passion. Which is precisely why, in vernacular language, makers and critics often refer to a play or a performance as 'a piece', *a part* that remains *apart*, that can never be placed with a system that would regulate it simply because it is forever on the move, again and again.

Collaboration

Nick and I frame our more-than-human performances with rivers, roads, sanitation plants, buckets, pebbles, paving slabs, weather and countless other things as collaborations. We do this while recognising the inequities and paradoxes of this framing. What, after all, is collaboration to water or a rock? More-than-human creative partnerships are always asymmetric. In casting them as our collaborators, we anthropomorphise the water, stones, drains and sidewalks with which we perform. We acknowledge the consequences and limits of doing so but venture that tactical anthropomorphising might be one of the few strategies we have for attuning to the other-than-human, and for unsettling the human exceptionalism that has led to climate change and environmental crisis. Describing the other-than-human participants in a performance as collaborators, we propose, acknowledge their agency and functions as a tactic to unsettle the anthropocentric hierarchies that persist in ideologies underpinned by Western humanism. Labelling the entangled agencies of more-than-human performance as collaboration makes a gesture towards communication beyond the human. While also, perhaps, simultaneously recognising its failure. Our more-than-human collaborative practice recognises the inherent animism of all matter. An animism that has been, as science philosopher Isabelle Stengers attests, derided, compartmentalised, erased and dismissed from 'proper' (Western) scientific and rationalist thinking (2012, 5–6). Our practice of collaboration, we believe, helps us to recognise this animism and alerts us, in the words of Stengers, citing David Abram (Abram 2010, 281) that 'our senses are not for detached cognition but for participation, for sharing the metaphoric capacity of things that lure us or that recede into inert availability as our manner of participation vanishes' (Stengers 2012, 8). As I gently kick a small stone along a city street, I am immersed in a more-than-human performance: pebble, foot, tarmac, granite kerbstone, shoe, ankle-joint, pothole and innumerable others all participating in *Rolling Stone Drift*. I notice subtle sonic variations as the stone skitters over concrete tiles or mudstone paving flags. I shift my weight, responding to undulations in the side-walk underfoot. I delight in the game: the skip and bounce of the pebble. Then it escapes, slipping between the slots of an iron drain cover. My participation in the performance is over until the next time.

Plague

Artaud's celebrated performance lecture of 1933, 'Theatre and the Plague' is crucial to climate thinking. In it, Artaud is concerned to situate the human within nature, blocking all escape routes from the earth. In doing so, he undermines, in advance, the dangerous logic inherent in eco-modernist ideas of rapture, the dreams that many technophiles have of exiting the planet, downloading the soul, and transcending the body. Artaud knows that vaccination provides a false perspective, lulling one into thinking that the human animal is immune, a scenario, which, as we have seen, so intensely and insanely, in our supposedly post-COVID world, only perpetuates destruction. To be vaccinated is to be made 'bullet proof', to lord it over the world, to put on body armour. To rethink Artaud's idea of plague, then, is to affirm theatre as a terrestrial act precisely because it is a creaturely act, a 'doing' that implicates spectators, leaving bodies exposed and vulnerable, able to be wounded by forces over which they have no control, and which they cannot see. Plague theatre does not simply convey a dark energetics, a type of aesthetic cruelty; crucially, it grounds that energetics and cruelty in the earth's own anarchy–its indifference to the fate of the *anthropos*. This from Artaud: 'We are not free, the sky can still fall on our heads. And that theatre has been created to teach us that *first of all*' (1958 [1936], 79; my emphasis).

Vigil

Theatre, as we all know, is related to the act of looking, a scopic medium predicated on the idea of the *theatron* (a place for seeing). But perhaps the sense of comforting distance inherent to the *theatron* needs to be challenged for a viable theatre ecology to emerge. Today, looking is not–cannot–be disinterested; rather, it needs to exist as a sensorium of emergency, a quickening of the eye, an organ brought down to earth. In the place of the *theatron*, a different matrix for the visual is needed, a technology of vigilance. To engage in a vigil is to be employed in an act of interested looking, to attend to the sick and dying; or to be like a lookout at sea, in the crow's nest, attuned to icebergs, reefs, stormy weather. Or, more theatrically, to be like the guards Marcellus, Bernardo and their anonymous companions–whom we meet on the ramparts, in the middle of a cold, dark night in Northern Europe, in *Hamlet.* These are the sentinels or 'vigilants' who witnessed the return of a ghost demanding justice, asking what it means '*to learn to live, finally*' (Derrida 1994, xvii), as Jacques Derrida parses the scene in *Spectres of Marx: The State of the Debt, the Work of Mourning & the New International.* Though he never intended it, the title of Derrida's text now seems so ecologically resonant, so presciently engaged with what Timothy Morton has famously termed 'the 'hyperobject' of climate change (2013)–a temporal weirdness, a return of the past in the future and *vice versa.* To be vigilant, then, is to participate in an impossible time, a spectral duration that splits the present, making it impossible but necessary. This is the time of our time, the time of the Anthropocene.

How, then, to transform a *theatron* into a vigil? And what might such a changed way of theatrical looking do? To be vigilant is not to remain apart from what you are witnessing; it is to be implicated, affected, contaminated. One thinks of Kris Verdonck's extinction trilogy that runs from 2008 to 2018: *End, Conversations (at the End of the World), Something (out of Nothing).* These are performances that insist on vigilance. Durational events that do not just show the falling apart, the weathering of bodies and things as they undergo time's impress. They demand that we, the audience, undergo that weathering too, and that we are able to recognise the stakes of that process as it is happening. That lexical torque, this shift towards vigilance, is an attempt to contest theatre's production of the god's eye view, the way of seeing that ecomodernists and techno-utopians use when they consider climate breakdown as something to manage and the earth as a supposedly stable object whose future can be predicted and measured, according to human ratios and desires.

Guddle

Guddle, or guddling, is a word Nick and I use to describe our more-than-human practice, particularly when working with water. 'Guddling About' is the title we give to the playful, small-scale actions, or micro-performances, which we devise with rivers, canals, sewage treatment works, puddles and other watery environments. Guddle is a Scots language word. It is a word that is in my everyday vocabulary and my memories. I can hear my mother's voice, telling me to 'stop guddling about'. Guddling situates me. It tethers me, loosely, to a place, time and culture. To guddle is to mess about. It is to act in an undirected, intuitive way. Guddling can be playful. It is a somatic practice, which requires attuning to an environment through touch, sight, sound and smell. A second definition of the word is to catch fish by hand, groping under banks and rocks in rivers where they lurk. Guddling is a watery word. It is onomatopoeic. Guddling sounds like the noise of a creek or stream running over stones or of water agitated by hand. As a noun (to be in a guddle), the word means to be in a muddle, a state of confusion or an entanglement. Guddling connotes the ludic, the sensory, a form of unassuming and improvisatory activity, a form of not knowing. Guddling might be the best we can do to live and practice in an era of climate emergency. To loop back to the very beginning of this chapter, guddling is also a word that underscores the need for a new lexicon for theatre ecology–one that creates a series of conjunctions between theory and practice, words and acts, performance and earth.

Note

1 Macarena Gómez-Barris explains that 'by addressing "humanity" as a whole', the Anthropocene veils the 'histories of racial thought and settler colonialism that are imposed upon categorizations of biodiversity, spaces where the biotechnologies of capitalism accelerate' (2017, 4).

References

Abram, David. 2010. *Becoming Animal: An Earthly Cosmology.* New York: Vintage.

Alaimo, Stacy. 2010. *Bodily Natures: Science, Environment and the Material Self.* Bloomington: Indiana University Press.

Artaud, Antonin. 1958 [1936]. 'The Theater and the Plague.' In *The Theater and Its Double.* Translated by M. Caroline Richards. New York: Grove Press.

Benjamin, Walter. 1999. *Selected Writings*, Volume 2, Part 2 1931–1934. M. W. Jennings, H. Eiland and G. Smith, eds. Translated by Rodney Livingstone and others. Cambridge, MA: Harvard University Press.

Bennett, Jane. 2005. 'The Agency of Assemblages and the North American Blackout.' *Public Culture* 17 (3), 445–65.

Connor, Steven. 2014. *Beckett, Modernism and the Material Imagination.* Cambridge: Cambridge University Press.

Derrida, Jacques. 1994 [1993]. *Spectres of Marx: The State of the Debt, the Work of Mourning & the New International.* Translated by Peggy Kamuf. London: Routledge.

Gómez-Barris, Macarena. 2017. *The Extractive Zone: Social Ecologies and Decolonial Perspectives.* Durham, NC: Duke University Press.

Haraway, Donna. 2016. *Staying with the Trouble: Making Kin in the Cthulucene.* Durham, NC: Duke University Press.

Latour, Bruno. 2018 [2017]. *Down to Earth: Politics in the New Climatic Regime.* Translated by Catherine Porter. Cambridge: Polity.

Morton, Timothy. 2013. *Hyperobjects: Philosophy and Ecology after the End of the World.* Minneapolis: University of Minnesota Press.

Nixon, Rob. 2011. *Slow Violence and the Environmentalism of the Poor.* Cambridge, MA: Harvard University Press.

Rancière, Jacques. 2004. *The Politics of Aesthetics: The Distribution of the Sensible.* Translated by Gabriel Rockill. London: Continuum.

Read, Alan. 2008. *Theatre, Intimacy & Engagement: The Last Human Venue*. Basingstoke: Palgrave Macmillan.

Schweitzer, Marlis and Joanne Zerdy. 2014. 'Introduction: Object Lessons.' In Marlis Schweitzer and Joanne Zerdy, eds, *Performing Objects and Theatrical Things*. Basingstoke: Palgrave Macmillan, 1–17.

Stengers, Isabelle. 2012. 'Reclaiming Animism.' *e-flux* (36), 1–10. https://www.e-flux.com/journal/36/61245/reclaiming-animism.

Werry, Margaret. 2020. 'Decolonising Theatre History: Ontological Alterity, Acting Objects, and What Theatre Studies Can Learn from Museums.' In Tracy Davis and Peter Marx, eds, *The Routledge Companion to Theatre and Performance Historiography*. London: Routledge, 206–26.

4

HOPE AND TRUST IN THE FUTURE

Human to Nonhuman in Australian Drama

Peta Tait

The climate change era brings hope into focus, even though it might seem a nebulous response. Theatre about climate change may or may not elicit hope, even though the performance form itself depends on the trust of an audience. In their summary of approaches to an interactive theatrical engagement between spectators and performers, Mireia Aragay and Enric Monforte recognise that claims for oscillations of feeling are underpinned by a 'relationship based on trust', a 'mutual trust' (2016, 16, 18). Trust becomes evident from the action of stepping into a performance space with others. Trust is less idealistic than hope, and it is a foundational premise of walking forward in performance—and in life (Tait 2022).

This chapter explores how hope and/or trust might encourage the forward momentum needed by humans and nonhuman animals (henceforth animals) within an uncertain future. Trust corresponds with the intensity of affect in its recent usage as something, 'filled with motion, vibratory motion, resonation' (Massumi 2002, 26). Trust underpins a body's impulse to move, so it need not be conscious. In contrast, hope is usually deemed an emotional feeling in Western culture, albeit a contested intellectual one (Averill 1996, 26–27). But trust can also be aligned with emotional feeling such as in 'affectionate trust' or the loss of trust through someone's anger (Oatley 2000, 95). Hope is harder to quantify than other emotional feelings such as fear and anger and, interestingly, with joy are easily identified across cultures (Dharwadker 2015, 1400–01). Since Charles Darwin (1999) claims animals also express joy, which suggests it has an evolutionary purpose, what insights might animals offer about trust and joy at a time when hope and hoping for the best seem unreliable?

As this discussion considers examples of trust and hope in Australian dramatic theatre, it contends that trust becomes a precondition of hope in human

DOI: 10.4324/9781003584445-6

resistance to destructive environmental practices and in patterns of human-animal relations. Trust is foundational to theatrical performance because audience participants trust that the unfolding circumstances will be beneficial – even joyful – rather than harmful. Cultural and species differences, however, merge to complicate processes of trust and trusting in the human-to-human scenarios in, *Namatjira: Written for the Namatjira Family (Aranda)*, by Scott Rankin and Trevor Jamieson, and in the human-to-animal exchanges in *Dogged* by Andrea James and Catherine Ryan.

A larger point is that because trust and hope reflect emotional responses to climate change, they need to be scrutinised. This is particularly so given scepticism about hope's capacity and the value of emphasising 'guilt' and 'collective guilt' about carbon emissions combined with 'culpability', and recognition of 'ecological grief' (Jensen 2019, 26, 113, 125, 141). In her comprehensive investigation of hope, Lesley Head (2016) contends that grieving and collective grief about ecological loss are all-pervasive and imbue practices deemed hopeful so that grief must be consciously acknowledged and endured for progress on climate change. Head's hope is not about optimism or even a feeling but manifests within a diverse range of cultural practices including behaviour and speech. While all these emotions encapsulate responses to harmful predictions of climate disaster and death, it is an absence of trust in shared human-animal ecologies in particular that prefigures perception(s) of harm. A twenty-first-century rise in human mistrust suggests a precursor to hopelessness.

Hope is often assumed to offer a viable antidote to despair over the slowness of action on climate change rather than prefiguring a companion of grief. Catriona McKinnon writes in this context: 'Hope aims at an objective which exists in the future, is valued by the hoper, and is desired' together with 'the belief that the future in which its objective exists' is possible (2014, 34). This hope requires belief. The danger with climate change is that scholarship shows 'that which is repeatedly asserted to be impossible can thereby become impossible' (McKinnon 2014, 45). Here, hope is justifiable and deliberate, although individualised, even though there is a presumption in Western culture that hope arises in shared circumstances and is potentially relational. Access to hope and its practices, however, needs some reconciliation with its contradictory history.

Christian Hope

Trust might be fundamental to human-to-human relationships, but its importance is often only recognised by its absence, while hope is often described and esteemed. Hope, however, has been studied in relation to Christian belief. James Averill writes that a history of hope in Western culture can be traced in religious writing, and accordingly, hope continues to be widely perceived in other people, even though it is not expressed in the same expressive way as, for example, love or anger (1996, 26–27). Hope is interpreted in metaphoric ways for its capacity to motivate and change thinking and behaviour. Therefore, hope is much more language dependent than trust, which can manifest through bodily action. The perception of either the affect of trust or an emotional feeling of hope is more complex than (basic) emotional feelings such as fear or joy, which are understood to arise in the rapid response of an orientated 'appraisal' of stimuli in the immediate surroundings (Parkinson et al. 2005). In distinguishing emotional feeling from expressive exchange, Averill, however, finds cultural difference in, for example, the way socially acquired hope in South Korea exists as a stable influence on a person, whereas hope is considered a subjective, fluctuating mode of coping in Western psychology (1996, 31). Hope might have expanded into a 'science' from its older status as a virtue (Gallagher and Lopez 2017; Mittleman 2009), but the interpretative language of hope needs a cultural context.

The Aranda writer-narrator, Trevor Jamieson, gives a humorous account of trust and mistrust in responses to Christianity in Scott Rankin's *Namatjira*. 'Trevor' who is named in the play script, explains that the effort to get 'Aranda mob' (First Nations central desert people, also spelt Arrernte) to convert to Christianity under the presumption it will benefit them and bring hope, involved the German Pastor missionary speaking Aranda words. But he uses the wrong words so the mob were singing the words for following a kangaroo or a quoll (small mammal) in hunting. The mob were singing: '*tracking to kill, I will spear Jesus*' until this mistake was realised. (Rankin 2012, 16, italics in original). It was changed to walking in front (of Jesus) and singing,

> I completely trust you, Jesus,
> To walk behind me and not kill me
> Even though you are carrying a very sharp spear.
> (Rankin 2012, 16)

The translation between cultural languages remains framed in the practice of spearing animals, according to First Nations traditional culture. Hunting gives trust particular interspecies significance in *Namatjira*, one intricately bound up with an understanding of a world in which a free-roaming animal would mistrust human proximity and intention, and would flee accordingly. Irony arises from language that puts community members in front of Jesus, to suggest they should mistrust what is happening and flee from Christian conversion. At that time, 'Trevor' explains, the underlying brutal reality of the mid-twentieth century included First Nations peoples in Australia being grouped with 'flora and fauna' by the government (Rankin 2012, 32).

While the Aranda peoples show trust in stepping forward, the imposition of Christianity typifies great cultural loss in a backwards regression under colonialism. The widely read *Namatjira* text is about the life of the famous Australian landscape artist, Albert Namatjira, his community and his extraordinary art (French 2002; Rankin 2018). The play also encompasses the history of race relations and environmental degradation in twentieth-century Australia. It outlines how the character of the artist, named 'Albert' in the play (henceforth Albert), and his family were forced off their traditional Country (lands) in the desert regions (Nations) surrounding Alice Springs. They were sent to live at the Hermannsburg Christian Mission. Yet Albert becomes Australia's most popular landscape painter with a celebratory and joyous depiction of gum trees and natural ecologies in his paintings. The play contains bilingual components and encompasses the close relationship to Country and belief in the non-hierarchical cohabitation of all species. Tradition allocates human custodianship of biodiverse places and protective association with animal species. The play describes how the dispossession of traditional lands led to severe damage in a short time as cattle were introduced by the colonial settlers and ate the native grasses and trampled down and polluted the waterholes, so that the kangaroos who fed on native grasses disappear, as did the grass seeds harvested for traditional flour (Rankin 2012, 22).

Imperial agricultural practices and imported animals typified the colonial occupation of land across Australia, which was bound up with Christianity. As First Nations peoples continue to resist ecological damage and the destruction of culture, language and identity, hope in the future might seem elusive. Struggling against patterns of genocide and ecocide, First Nations peoples maintain their law, knowledges and ongoing care for Country as they fight for land rights (Langton and Corn 2023).

Even though Albert's people are forced off their Country, the Aranda mob in *Namatjira* maintain caring responsibilities for Country and for the animal species allotted through kinship connections. Albert explains to his non-Indigenous collaborator and artist friend, Rex, that there are restrictions as to where he, Albert, can look to paint, because some places are women's places, and some are not accessible to his kinship lineage. For example, where 'lizard men climb sheer rocks' cannot be painted (Rankin 2012, 25). But Albert can paint a 'hopping kangaroo' (2012, 29). His landscapes become highly prized and he financially supports his 600-strong community as he becomes a national celebrity and is given a standing ovation in a state theatre auditorium (Rankin 2012, 41). One of Albert's paintings is presented to Queen Elizabeth II circa 1953. Contradictions abound as Albert and his community continue to be racially persecuted, and Trevor describes how Albert travels from 'Government House to Government Dump' and prison (Rankin 2012, 45). Albert's prominence in his community means that he gets blamed for their access to alcohol, and he is falsely accused of attacking a woman, even though he was not there. The innocent Albert is accused of 'a serious betrayal of the trust' placed in him and sentenced to six months in prison to set an example to his community and he dies having been 'used, abused and then abandoned' in a Jesus-like betrayal (2012, 44, 46). Albert seems despairing as he asks, 'Who am I?' (Rankin 2012, 43).

In *Namatjira*, there is a hope that Australian society will change after soldiers including First Nations soldiers, return from World Wars. There is hope that Albert's achievement will embody the changed social and economic status of dispossessed First Nations peoples, given the success of his painting. These hopes are not realised in Albert's lifetime. The hope for a better future demanded by Christianity is repeatedly betrayed by a racist white Australia and its land practices. Meanwhile, where possible, Namatjira's 'Aranda mob' continue to covertly return to and look after their Country and its species. First Nations peoples continue to practice trust in the Country that they share with other species.

Namatjira suggests that settler society betrays hope for the future including hope for ecological protection. Trust, however, remains in the physiological processes of an inexorable forward motion. Yet the human right to hope is important, and it can be aligned with 'atonement', and it is studied in the context of prisons (Trotter 2022, 2). Atonement can also encourage land rights and ecological restoration. In a discussion of hope in relation to the Indigenous North American Crow nation that also experienced colonising devastation, Jonathan Lear considers the cultural context that connected hope to courage, so it becomes radical when 'it is directed toward a future goodness that transcends the current ability to understand what it is' (2008, 103). McKinnon's future objective might be unforeseeable, but hope can be accompanied by 'dignity' and, significantly, it relies on social relations (Trotter 2022, 2). It is this type of relational hope that promises to oppose the oppression that diminishes hope. Lisa Kretz argues for 'a relational conceptualization of hope where hope is only possible in supportive relations' (2015, 136). Hope that emerges within groups who oppose an oppressive history that induces hopelessness, expands on affects of trust.

Dingo Trust

The creation stories about 'caterpillar dreaming' and the 'devil dog' dingo in *Namatjira* are ostensibly suppressed under Christian beliefs (Rankin 2012, 43, 44). The dingo encapsulates continuity across time while it also embodies the late twentieth-century fight to preserve healthy ecological systems in Australia. The dingo is a native canine species that is genetically distinct from dogs, and while the dingo is associated with cultural identity in some First Nations cultures in Australia, dingoes were hunted and much maligned in historic settler culture. In summarising the dominant approaches to Australian native animals, Rick De Vos explains that 'colonialism and settlement shape and mark out shared landscapes and habitats' privileging human-centric Western epistemologies to produce 'exclusions and inclusions' that emphasise some 'forms of life while obscuring others' (2023, 4). These binary separations are ecologically damaging, and while central to some First Nations belief systems, the dingo species stands for a category of exclusion within settler culture.

Dingoes are among the native animals routinely shot and poisoned on private land, although a select few are held in captivity for 'pleasure and profit' (Boyde 2014, 2). The dingo was once important to the ecological balance but maligned by settler fear that dingoes will attack farm livestock, and wild dogs and dingoes are lumped together. The species distinction between dingoes and wild dogs is central to the bleak, timeless world of *Dogged* by Andrea James and Catherine Ryan which depicts how both are indiscriminately shot by hunters. (It implicates rural land management and in 2024, the state government of Victoria reapproves the culling of dingoes in the southwest region.) The hungry character, Dingo, runs through the bush desperately (doggedly) searching for her babies who ran away when Dingo's male partner was shot by hunters. She encounters the characters of a Woman accompanied by Dog on her hunt. As Dingo points out, hunters are coming into her territory—which was once cohabited with First peoples (James and Ryan 2021, 20).

Yorta Yorta/Gunaikurnai writer, Andrea James, presents a dramatic chorus of dingoes narrating events in her *Yangali Yangali* play about land rights from 2003 (Tait, Varney and Stevens 2025), and in *Dogged*, she co-writes Dingo and Dog as full characters personified by human actors. It could be said to be part of a twenty-first-century literary genre that is literally giving voice to the concerns of other species and their future survival on behalf of human-damaged ecological systems. Anthropomorphic characterisation also reflects the proposition that humans become concerned about the plight of nonhuman species and their habitats when they are emotionally engaged. At the beginning of *Dogged*, Dingo watches the approaching Woman and Dog listening to heartbeats and mistrusts them. But Dingo is hungry and moves forward and Woman offers food, expressing fascination and respect. Dingo accepts Woman's offers, and the lessening of Dingo's mistrust is shown through her behavioural action of staying by the fire. Woman, however, is dependent on Dog for companionship and to hunt wild dogs—seemingly for a living. Subsequently, Dog joins the cautious Dingo in a rabbit hunt. When the guns of wild dog hunters are heard across the valley, Woman says, 'Another stinking boys' weekend killing party' with their 'savage crossbreed dogs' and spotlights (2021, 23, 27). She is angry about the indiscriminate brutality of masculine hunting and the play implies that Woman trusts Dingo more than her own species.

Dingo interrupts their exchanges with her joyfulness at the arrival of migratory moths who are 'beautiful' and 'magic' (2021, 23, 24). She points to environmental loss and food scarcity and warns Dog about the poisoned bait that he cannot resist and will later kill him. Dingo offers a nonhuman perspective about human behaviour in the natural world saying.

This damning indictment against settler hunters is followed by historical account of a 'Black War' with First Nations peoples, and Dingo also repeatedly reminds Woman that they are on 'Brabralung Country' (2021, 47).

> 'You crash in and tear us apart, /Like insatiable beasts ripping flesh from bones. And you dare to call us wild?' (2021, 43).

Dogged highlights mistrust as humans strive to dominate and kill. Paradoxically, the humans are completely dependent on the physical capacities of domesticated dogs for hunting. While Dingo conveys desperation about the loss of her babies, Dog cannot escape submission to the Woman who chains him up with cruel affection (2021, 21–22). Dog displays what Charles Darwin identifies as a recognisable 'dejection' in bodily signs of a dog, but then again Darwin calls it 'hopeless dejection' (1999, 62). In several Australian First Nations creation beliefs, the human is interchangeable with the nonhuman dingo (and quoll) and they share the same emotions and food (Lennox 2023). They might have trust in common, but do they share hope?

In his approach to the emotional phenomenology of hope and depression in humans, Matthew Ratcliffe (2013) connects mistrust in the world to loss of hopes which he distinguishes from the loss of hope. He explains that it is necessary 'to distinguish forms of hope' that are intentional, and can be either passive or active, from those that arise with 'belief, desire and expectation', that is, from future objectives (Ratcliffe 2013, 598). He recognises that there are social limits in the material conditions preventing individuals from achieving their personal hopes. But Ratcliffe argues that 'pre-intentional hope' is a bodily feeling and orientation that aligns with trust that there is a future (2013, 600–01, 604). Therefore, the loss of hope is felt as an absence of forward momentum. In humans, it manifests with severe depression and no belief in the future and 'loss of trust in the world' (2013, 612).

Even in the dystopian world of *Dogged*, Dingo displays trust in her bush environment as she runs forward looking for her pups. One baby dingo does appear towards the play's end as Dingo—emblematic of ecological loss—lies dying. Yet the baby dingo survives. De Vos argues that native 'nonhuman animals have displayed resilience, resistance and persistence' to colonialism and an ongoing disruptive capacity (2023, 5). Nonhuman animal presence can disrupt the dominant settler paradigms that induce mistrust and hopelessness through ecological loss.

Dingoes coexisted with First Nations peoples for at least 5,000 years with a capacity to lead them to water and food, and some lived near human groups. They are part of the whole environment that is integrated into ontological belief (Thurow 2018). In her collation of stories about dingoes, Rowena Lennox finds that they become associated with the creation of the material world giving knowledge to the living, and in some instances, as a type of devil dog who eats and regurgitates people in an ecological cycle (2023, 58, 61). Dingoes have philosophical significance as creation beings called, Ankotarinja and Erintja Ngoolya in Aranda-speaking Country in the Central Desert (Lennox 2023, 64). Lennox (2023) charts how dingoes continue to unsettle colonising frameworks, describing misperceptions ranging from the famous court trial of Lindy Chamberlain wrongly accused of infanticide when a dingo actually took her baby, to misbehaving tourists interacting to their detriment with the free living dingoes on Fraser Island. Lennox compares settler condemnation to the incomplete mission accounts that nonetheless reveal how the dingo has been associated with red and redness in traditional culture and possibly stands for 'joy and happiness' and the benefits of fire (2023, 64–65). The dingo that is free to roam in the environment is associated with joy.

Moving Forward

Charles Darwin finds joy evident in both humans and animals and especially in the expression of dogs through 'bounding' exuberant action, which is often 'purposeless movement' (1999, 80). Joy is communicable through spontaneous movement and behaviour. Antonio Damasio evaluates whether human joy 'signifies equilibrium for the organism' and joyousness means 'a greater ease in the capacity to act', although as a heightened experience, joy is linked to the opposite condition of sorrow in Damasio's interpretation of Spinoza (2003, 137–38). Joy seems to facilitate action and the science suggests joy leads to 'flourishing' in life (Damasio 2003, 285). As an emotional experience, joy prefigures an intense delight for a short time rather than a sustained condition. This suggests an experience that needs to be refreshed. Dingo in *Dogged* shows how ecological joy can alleviate and temporarily divert even mistrustful responses.

In the example of theatrical performance, the evocation of hope and/or joy can be hard to quantify within the subjective impact. While Aragay and Monforte acknowledge that interactive and immersive theatrical performance can involve ambivalence that must be offset by a performer's questioning concern for an audience, they recognise that claims made for 'connectedness', intimacy and spectator agency can be difficult to verify (2016, 7–8). The elucidation of relational hope requires linguistic communication whereas joy is recognisable through its facial and bodily expression but trust at least can be presumed by spectator attendance.

The dramatic examples here suggest that the trust arising from moving through the natural environment might inspire moments of joy. Yet the destruction of traditional knowledge and species and the loss of Country would seem to prefigure hopelessness. However, Trevor describes the artist, Albert in *Namatjira* producing artwork that conveys, 'Where the ghost and the rivergum glow like white giants, in sun harsh against the distant mauve horizon', to evoke the wonder of his mother's Country (Rankin 2012, 31, 36). Albert is inspired by the ecological world and his art invites nation-wide enjoyment of the landscapes that he knows are shared with other species and have been entrusted to human care. The drama reveals that Albert's paintings communicate profound environmental knowledges.

Dingo experiences moments of joy in *Dogged* even as her freedom and roaming are under threat. Her searching contains moments when: 'Eucalypt leaves gloss your feet, nostrils flare, Smells. Smells. Everywhere' (James and Ryan 2021, 15). Dingo cannot be separated from her environment so that pleasurable smells interrupt her fraught searching.

While the reliability of sustained hope for the future can be questioned, trust relies on the sensory body's capacity for engagement and onward movement. It does not emulate the belief required for hope which may or may not deliver action. Instead, trust in the future emerges from the incremental forward momentum of individuals and groups of all species and especially those undertaking urgently needed action on climate change. Let us hope we can continue to be inspired by ecological joyfulness.

References

Aragay, Mireia and Enric Monforte. 2016. 'Introduction: Theatre and Spectatorship – Meditations on Participation, Agency and Trust.' *JCDE* 4 (1), 3–20.

Averill, James. 1996. 'Intellectual Emotions.' In Rom Harré and W. Gerrod Parrott, eds, *The Emotions: Social, Cultural and Biological Dimensions*. London: Sage, 24–38.

Boyde, Melissa. 2014. 'Introduction.' In Melissa Boyde, ed., *Captured: The Animal Within Culture*. Basingstoke: Palgrave Macmillan, 1–8.

Damasio, Antonio. 2003. *Looking for Spinoza: Joy, Sorrow and the Feeling Brain*. Orlando, FL: Harcourt.

Darwin, Charles. 1999. *The Expression of the Emotions in Man and Animals*. London: Fontana Press.

De Vos, Rick. 2023. 'Introduction: Unsettling Subjects.' In Rick De Vos, ed., *Decolonising Animals*. Sydney: Sydney University Press, 1–17.

Dharwadker. Vinay. 2015. 'Emotion in Motion: The *Natyashastra*, Darwin and Affect Theory.' *PLMA* 130 (5), 1381–404.

French, Alison. 2002. *Seeing the Centre: The Art of Albert Namatjira, 1902–1959*. Canberra: National Gallery of Australia.

Gallagher, Matthew W. and Shane J. Lopez, eds. 2017. *The Oxford Handbook of Hope*. Oxford: Oxford UP.

Head, Lesley. 2016. *Hope and Grief in the Anthropocene*. London: Routledge.

James, Andrea. 2003. *Yanagali, Yanagalii!* Sydney: Currency Press.

James, Andrea and Catherine Ryan. 2021. *Dogged*. Sydney: Currency Press.

Jensen, Tim. 2019. *Ecologies of Guilt in Environmental Rhetoric*. Cham: Springer/Palgrave Macmillan.

Kretz, Lisa. 2015. 'Singing Hope's Praises: A Defense of the Virtue of Hope for Environmental Action.' In Andrew T. Brei, ed., *Ecology, Ethics and Hope*. Lanham, PA: Rowman & Littlefield Publishers, 133–50.

Langon, Marcia and Aaron Corn. 2023. *Law: The Way of the Ancestors*. Melbourne: Thames and Hudson.

Lear Jonathan. 2008. *Radical Hope*. Cambridge: Harvard University Press.

Lennox, Rowena. 2023. '"Red I am": Names for Dingoes in Science and Story.' In Rick De Vos, ed., *Decolonising Animals*. Sydney: Sydney University Press, 39–73.

Massumi, Brian. 2002. *Parables for the Virtual*. Durham: Duke University Press.

McKinnon, Catriona. 2014. 'Climate Change: Against Despair.' *Ethics and the Environment* 19 (1), 31–48.

Mittleman, Alan. 2009. *Hope in a Democratic Age: Philosophy, Religion and Political Theory*. Oxford: Oxford University Press.

Oatley, Keith. 2000. 'The Sentiments and Beliefs of Distributed Cognition'. In Nico Frijda, Antony S. R. Manstead and Sacha Bem, eds, *Emotions and Beliefs*. Cambridge: Cambridge University Press, 78–107.

Parkinson, Brian, Agneta H. Fischer, and Antony S. R. Manstead. 2005. *Emotion in Social Relations: Cultural, Group, and Interpersonal Processes*. New York: Psychology Press.

Rankin, Scott (with Trevor Jamieson). 2012. *Namatjira: Written for the Namatjira Family (Aranda)*. Sydney: Currency Press.

Rankin, Scott. 2018. 'Namatjira Project: What Is It That We Are Not Seeing?' *Meanjin* 77 (2), 52–58.

Ratcliffe, Matthew. 2013. 'What Is It To Lose Hope?' *Phenomenological Cognitive Science* 12, 597–614.

Tait, Peta. 2022. *Forms of Emotion: Human to Nonhuman in Drama, Theatre and Performance*. London: Routledge.

Tait, Peta, Varney, Denise, Stevens, Lara. 2025. *Ecology and Climate in Theatre and Australian Performance*. Cambridge: Cambridge University Press.

Thurow, Suzanne. 2018. '"Namatjira": Beyond the Script – Visual and Performative Aesthetics as Conduits for the Communication of Western Aranda Ontology.' *Australian Drama Studies* 73, 130–59.

Trotter, Sarah. 2022. 'Hope's Relations: A Theory of the 'Right to Hope' In European Human Rights Law.' *Human Rights Law Review* 22, 1–21.

Provocations in the Anthropocene

More-Than-Human

5

THE SEA IS ON FIRE

Machinic Crustaceans and Ecological Promises

Jen Parker-Starbuck

> "And do you know", I added, "what has been the result since men have almost entirely annihilated this useful race? That the putrefied weeds have poisoned the air, and the poisoned air causes the yellow fever, that desolates these beautiful countries. Enormous vegetations are multiplied under the torrid seas, and the evil is irresistibly developed from the mouth of the Rio de la Plata to Florida. If we are to believe Toussenel, this plague is nothing to what it would be if the seas were cleared of whales and deals. Then, infested with poulps, medusae, and cuttle-fish, they would have become immense centres of infection..."
>
> *(Verne 2020, 215)*

Within Jules Verne's epic novel, between the detailed descriptions and taxonomies of sea creatures and seaweed, and amidst the grand-journey storytelling, this passage (written between 1869 and 1870) introduces an environmental warning: If large numbers of whales and seals are eradicated from the seas, oceans will be 'centres of infection' (Verne 2020, 215). While reading this novel during the 2020 Coronavirus pandemic, I was struck by the resonance of this passage in what is being called our 'climate crisis'.[1] The novel is better known for its futuristic technological submarine, the Nautilus, its invocation of the underwater world of Atlantis, its elusive Captain Nemo and its in-depth depictions of underwater life, but this ecological warning echoes through any reading of this novel today. The observations in the novel follow an interest in Darwin's marine life collections on the HMS Beagle in the 1830s, and the early understandings of aquatic life largely at the surface of the ocean. However, technological advancements continue to provide 'deeper' knowledge, which has contributed both to

DOI: 10.4324/9781003584445-8

human understanding, and destruction of, oceanic life. Verne fictionalises early findings in the novel. In, 'From Darwin to the Census of Marine Life: Marine Biology as Big Science', author Niki Vermeulen explains a history of some of these technologies:

> For a long time it was thought that life could only be found there [near the surface and] at the ocean surface, as the absence of light, low temperatures and the density of water in the deep ocean was assumed to prevent life. However, these ideas slowly changed with the development of technologies that made the deep ocean visible.
>
> *(Vermeulen 2013, 3)*

Vermeulen goes on to explain the development of these technologies from 'a church bell and an ear trumpet' to 'explosions' that would be used by the Atlantic Telegraphy Company to lay down the first trans-oceanic cable in 1858 (Vermeulen 2013, 3). Understanding the depth of the ocean has also prompted further technological developments such as submarine pipelines and offshore drilling. Discovery has led to destruction. In July of 2021, the waters in the Gulf of Mexico were 'on fire'.[2] An underwater pipeline burst, and the leakage caused what was described as an 'eye of fire' on the water. The petroleum company responsible, Pemex, denied any environmental damage, while Greenpeace argued that the damage was caused by gas leakage (and pointed out that had it not caught on fire we would not know how much gas had been leaking or for how long). In an article for *Climate Home News*, Greenpeace representative Pablo Ramirez explains that:

> The gas can rapidly penetrate the bodies of fish, doing direct damage to gills, skin, chemoreceptors and eyes, and filling up the gas bladder, making the fish unable to control its buoyancy … Shellfish are also killed by exposure to gas. If fish are exposed to gas concentrations of 1mg per litre they show signs of acute poisoning within 20 minutes and die within two days.
>
> *(Gerretsen 2021)*

Of course, this is not the first pipeline explosion, and nor will, I assume, it be the last. The list of pipeline accidents just on Wikipedia is daunting. This particular incident reportedly killed five people, and a very similar disaster occurred in 2015, when a fire broke out on, again, a Pemex oil platform, killing four humans (the nonhuman life is not reported). I was struck recently by the important inclusion in Denise Varney and Lara Stevens's essay 'The Climate Siren: Hanna Cormick's *The Mermaid*', when they described the 'hundreds of thousands of dead, colourless fish floating on the surface of the Murray-Darling … river system' due to neocolonial irrigation practices (2022, 114). Nonhumans are too often neglected in these disasters. It is also important to remember the Deepwater oil spill in 2010, thought to be the largest oil spill and one of the worst environmental disasters in US history. There, images of oil-soaked birds at least captured the public's attention. What is known is that, in climate disaster scenarios, the death and deformation of marine and bird life is immeasurable.

The human-technological-nonhuman relationship is complex. On the one hand, technologies have made it possible for humans to better know the world around them; machines permeate human lives and are relied upon for all manner of use and connection. But can they also be deployed as intermediary sites of mutual exchange or balance? Verne's explorers presciently walk the ocean floor and offer readers a detailed and colourful portrait of the wonders found there—wonders almost none of them will have ever experienced. The emerging technologies in Verne's novel contribute to the quest for knowledge and understanding of nonhuman sea life, but also are framed within a familiar narrative of killing and destruction. In the text, the narrator, Professor Aronnax and his companions Ned Land and Conseil are captured and find themselves on the Nautilus after an expedition to kill a giant never-before-seen creature that turns out to be a submarine. Among Verne's inventions in this novel is also a special gun that produced electric volts through its glass and steel bullets (akin to a taser) to kill the various animals. These are not technologies of mutual exchange.

It is technology's role in human relationships to nonhumans that I am concerned with in this chapter. While I largely feature the fantastical mechanical creatures found at Les Machines des L'île in Nantes, France, a 'theme park' of creatures set in a former shipyard, I want also to begin weaving in the conflicting role technology has in 'exploration' and environmental concerns. As expressed in the play, *Yellowfin*, by Marek Horn, 'when the fish went' synthetic cloned fish were developed to replace them (2021, 50). Human and nonhuman forces are both mediated through the very technologies—like the Nautilus or the underground leaking pipeline—that also keep them apart.

Les Machines

I began with Verne's novel because Nantes, France is also the birthplace of Jules Verne, and Les Machines' website proclaims that it is 'the only place where you'll find Jules Verne's "Invented Worlds", the mechanical universe of Leonardo da Vinci, and Nantes' industrial history, all on the exceptional site of the city's former shipyards' (Les Machines des L'île 2025). In considering Verne's 'Invented Worlds', it is the submarine that he most predicted, then a fantastical invention, with features well beyond the working models of the time. For Verne, the technologies allow for a deeper understanding of the marine world, one that remained removed from most humans. The marine life—from descriptions of left and right 'handed' shells to the size, shape and colour of various fish—is portrayed fairly accurately, if perhaps exaggerated or seemingly invented for the fiction reader in the 1860s. Note, as an example, this wonderful, descriptive passage:

> I distinguished some green labre, the banded mullet, marked by a double line of black; the round-tailed goby, of a white colour, with violet spots on the back; the Japanese scombrus, a beautiful mackerel of these seas, with a blue body and silvery head; the brilliant azurors, whose name alone defies description … Japanese salamanders, spider lampreys, serpents six feet long, with eyes small and lively, and a huge mouth bristling with teeth; with many other species.
>
> *(Verne 2020, 60)*

The detailed descriptions found throughout the book introduce readers to the findings of explorations, collections and cataloguing practices of the time.[3] The book *Animalia: An Anti-Imperial Bestiary for Our Times* acknowledges and attempts to disrupt histories of British imperialist use of animal classification in settler colonial destruction, and in this sense, Verne's novel might be considered as the authors assert, a kind of writing making its way into the hands of lay readers that 'blurred the lines between science, expertise, and popular culture' (Burton and Mawani 2020, 4) (Figures 5.1 and 5.2).

FIGURE 5.1 Le Carrousel des Mondes Marins. Les Machines de l'île. Nantes (Loire-Atlantique) © Franck Tomps _ LVAN.

FIGURE 5.2 Le Carrousel des Mondes Marins. Les Machines de l'île. Nantes (Loire-Atlantique) © Romain Peneau _ LVAN.

Les Machines des L'île's 'invented worlds' are not concerned with the submarine technologies as much as nonhuman animals themselves—from a spectacular walking elephant and slow-moving sloth to a carousel of Verne-esque sea creatures. Developed by artists François Delarozière and Pierre Orefice, Les Machine is a fantastical, Steampunk styled project where spectators can marvel at, as well as ride, giant machinic constructions of animals, insects and sea creatures.[4] After entering the main hall in which half is dedicated to a grouping of machinic creatures dotted around the warehouse space, groups are toured through and a lucky participant is able to ride the giant spider or flying heron. The other half of this space is a huge workshop in which their newest projects are under construction, and from there, spectators can purchase tickets to ride the 12 metre tall elephant across the plaza ending at the sea creature carousel, 'The Carrousel des Mondes Marines'. It is here where Verne's world is inverted. Within this structure, across three levels, you can ride the giant sea-life crustacean carousel, choosing between, for example, a crab, manta ray, giant squid, a large shell, or even descend underground in a submarine. The company's press kit parallels Verne's depth explorations:

> The visitors are spectators of strange and disturbing marine creatures that revolve in a huge room built on three levels. They discover the sea like never before, from seabed and abyss to sea surface.

> **In the seabed**, the public discover the Giant Crab, the Reverse Propulsion Squid, the exploring Machine that dives deep into the machine room, or the Bathyscaphe that climbs up the central mast… and the last one: the Box-fish. A total of 14 elements fixed onto a rotating plateau.
>
> **Into the abyss**, on the second level, 6 elements hang 5 metres above the seabeds: the Deep-Sea lanternfish, the Manta Ray or else the Pirate fish… Only visitors over 1m35 (4.5 ft.) are accepted aboard this level.
>
> **On the sea surface**, on the third level, a rotating plateau, protected by a Big Top, sweeps away boats, marine harnesses and other flying Fish, a storm Boat, Nutshells and Jellyfish, in a raging whirlpool of 24 big mechanical waves.
>
> *(Les Machines des L'île 2019, 8)*

In contrast to Verne's characters, who watch the array of sea life from within technological means/submarine, or while walking on the ocean floor in a diving costume, here visitors experience the sea creatures *as* the technological—machinic moving creatures to climb upon, operate, and marvel at (Figure 5.3).

FIGURE 5.3 Giant Squid at Les Machines des L'île. Photo by author, Jen Parker-Starbuck.

These machinic animals stand in for their real and fictitious counterparts, enacting a physical shift away from the viewing position of the spectator/observer and towards an active and engaged experiential position. I am interested in this technological reversal in a moment when nonhuman animals are so mediated and technologised and am reflecting here on Akira Mizuta Lippit's consideration of the merging of nonhuman animals and technologies that ultimately turn technology into a 'species' (2000, 187). But whether these technologies contribute to human understanding of these other lives is questionable. What is at risk when the nonhuman animal is technologised? Are there possibilities of resistance for these techno-animals? And can they have their own agency of sorts?

The 'squibs' as they are called in the play *Yellowfin*, by Marek Horn, are definitely a new kind of technological species. In this play, a courtroom drama in which a three person Senate committee is questioning a Mr Calantini about an alleged contaminated fish product, we learn that the fish have gone and what's more, they have disappeared overnight. This is a future world in which not only have the fish gone, but also there has clearly been catastrophic climate change. The character Marianne asks:

> How did the fish go?
> Where did the fish go?
> Why did the fish go?
> The three great questions of our age.
>
> *(Horn 2021, 66)*

The circumstances of the disappearance and the condition of the world is never clear, but the play (set in the United States) alludes to flooding, flooding that has decimated lands, as we see in this exchange:

Roy: … And the Dutch. I miss them too.
Stephen: The Dutch aren't gone.
Roy: Yeah I know.
I know that.
But I miss their tulips.
All those beautiful tulips….
Stephen: The land was too flat … so they had to go.
The tulips that is, not the Dutch.
…
Roy: Their houses floated, so they lived….
The English didn't float.
They didn't build the new houses so they didn't float (Horn 2021, 76–77).

This got a laugh in the London Southwark theatre when I saw it, but there is a seriousness to this play; its emergence at the same time of the COP26 (UN Climate Change Conference) in Glasgow is perhaps not coincidental. This is a world of flooding (a reality for the United Kingdom in climate crisis), in which the fish have gone (hearkening back to Verne's warnings), and there is a feeding frenzy about where to find the last few remaining tins of tuna to attempt to extract DNA and find answers about the disappearance of the fish. The play is about the climate and food and food production, and much time is spent trying to describe the difference between the 'spongy' squib and the flaking qualities of tuna (to one member of the committee who has never eaten actual fish). The play, like Verne's novel, emphasises a human need to consume the nonhuman, in this play quite literally. In this passage, Mr Calantini is asked to 'speak to flakage', and his reply beautifully captures for me the tension between humans and nonhuman animals:

CALANTINI:
Fish … y'see
Real fish, it
… *fragments*
It …*crumbles*
It breaks on the tongue like a wave on the shore
It feels like
Work
It feels like
Nature
And
Power
And
It feels like a thing … that has lived
Beat.
I mean, don't get me wrong, our fish live
The squibs are alive,
But it's a half-life,
It can't compare to
An
Animal,
A
Beast,
That has
Ripped
Through currents
And
Torn

> Through storms
> And through salt and pain and cold and space …
> And all that … *toil*
> And
> Force
> And
> *Instinct* …
> It *bleeds* in, I guess, in a sort of
> Chemical way.
>
> *(Horn 2021, 31–32)*

Reflecting the human admiration for animal majesty, this passage exalts fish, this force of nature (remember it is in their absence, and yet framed as domination and consumption), and the sense that its instinct and force and toil 'bleed in … in a chemical way' (32). No replica could capture this essence—this is his point. And yet, we turn to animal replicas for food, such as these squib, but also in robotics, intended for warfare, in entertainment and much more.

In sites/artworks/creations like Les Machines, the creatures connect participants to nature, to animal movement, to the sea. For adults and children alike, the company's website suggests that:

> There's no separation between the adult world and the world of children. Parents are not only accompanying, they are also concerned with the awakening of their children, and they share their trip, and discover behind the animal the machine in motion.
>
> *(Les Machines des L'île 2025)*

There is an attraction to these machines to be sure, and to the craft in their creation. Yet I am uncertain whether riding on a giant manta ray will increase 'awareness' later in life towards the environmental challenges the seas face; however, I have argued that an embodied experience may contribute towards a more *attuned* relationship between humans and nonhumans. Sitting astride a sea turtle with articulated joints and moving its head side-to-side will most definitely not lead to direct environmental change, but it may offer an opportunity of accretion learning about these creatures, 'swimming' beside them without harm, and producing an exploration into another world in the same way the Nautilus offered a technological means to explore the actual animals.

Exploring Technologies

I entered the Carrousel des Mondes Marines like a child in a candy store. First walking around each of the different levels and scoping the different creatures.

I bought a ticket to ride two of the creatures and wondered if I would recreate childhood experiences of riding a carousel horse, holding the reins and softly clucking to the horse as we moved up and down around and around. Of course, these mechanical creatures are of animals not intended for riding, so there was an unexpected sense of adventure. Strolling in wonderment, perhaps as Professor Arronax might have aboard the Nautilus, I selected the large giant squid, the creature that produced fear and revulsion in Arronax and his companions:

> Before my eyes was a horrible monster, worthy to figure in the legends of the marvellous. It was an immense cuttle-fish, being eight yards long. It swam crossways in the direction of the Nautilus with great speed, watching us with its enormous staring green eyes. Its eight arms, or rather feet, fixed to its head, that have given the name of cephalopod to these animals, were twice as long as its body, and were twisted like the furies' hair… The monster's mouth, a horned beak like a parrot's, opened and shut vertically… What a freak of nature, a bird's beak on a mollusc! Its spindle-like body formed a fleshy mass that might weigh 4000 to 5000 lbs; the varying colour changing with great rapidity, according to the irritation of the animal.
>
> *(Verne 2020, 220)*

I include this detailed description to note especially how it almost turns the creature into a technology, a 'freak of nature' with magical hybrid proportions and the ability to change colour.[5] The narrative was about fear and death, and after a crew member is killed by the giant creature, a battle soon ensues between the remaining crew and a school of these 'monsters', who are largely slaughtered, but leaving readers resonating with the reminder of the power and complexity of these underwater creatures.

In my own encounter, I was much less afraid, though equally as fascinated, but by the crafting of the mechanical squid. Beautifully carved with many moving pieces and strange gears protruding from its head, I desired to see how it moved, to ride with it and rather than conquer it, to work together as we spun around on the carousel. However, after my adventure, I can conclude that while aesthetically beautiful to behold, it was not as exciting or interspecies (nonhuman animal or technology) an experience as I had hoped. I laboured to make all the tentacles and eyes move and felt a need to keep it moving, to perform with it on our circular journey. I found myself, perhaps like Nemo's crew, trying to control or tame the beast rather than being at one with it, or just admiring it from afar (Figure 5.4).

In the absence of the living animal, there can be no ethical encounter—the machine replaces and removes the animal from display, use or service. The absenting of the animal through technological means may be to protect the abuse of nonhuman animals (in zoos or parks like Les Machines, or the removal of animals in circuses, for example), but is also a physical absenting, such as the current and

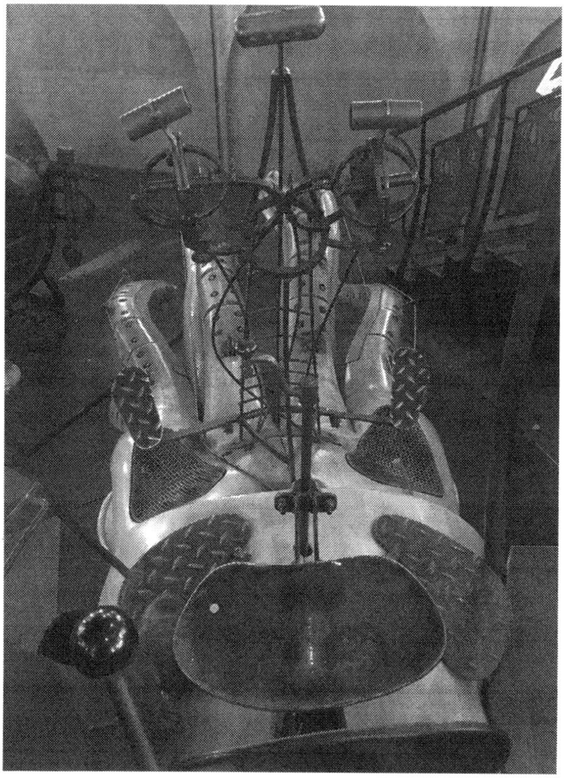

FIGURE 5.4 Giant Squid at Les Machines des L'île. Photo by author, Jen Parker-Starbuck.

continuing damage done by the leaking gas from underwater pipelines, conditions that led to the squib of *Yellowfin.* 'Technologised animality', as I am calling it, removes the face and the gaze of the nonhuman animal, beings to be accountable to, and replaces them with representations that erase the sense of responsibility to these creatures. Yet I am also attempting to analyse the possibilities of mechanised encounters that might have the potential to produce an awareness of responsibility, or kinship, or ethics. In riding the mechanised squid I was forced to contemplate my romanticisation of what this could be and instead had to work uncomfortably to force it to move. Perhaps this was the point; squids should not be forced to be at the service of humans. As this triangular human-nonhuman animal-technological encounter began to crack apart in my mind, it produced a tension that drew me more committed to, let's call it, a 'squid ethic'. For me, a resolve to let these creatures be.

Unlike my whispering to a carousel horse, or my ride on a mechanical squid, my encounters with living horses were much more affective. But for a child unable to be in a socioeconomic or rural context that might allow for an encounter with the live animal, perhaps sitting astride a carousel may still offer a connection, a desire. For me, my mechanised encounter drew me to Verne's novel, it has sent me down several rabbit, or should I say cephalopod-ean holes, it has made me want to save the oceans and their inhabitants.[6] In my embodied encounter with the mechanical squid, I knew I was a human atop this monumental creature, in an experience I would never have, nor should I have, in real life. I continue to search for embodied experiences, such as my current research into Virtual Reality experiences in which humans can be in the position of having an animal body or movement, so I can investigate how movement *as* nonhuman animal is different from movement *with* in relation to empathy, connectivity and animal agency. I was most certainly not 'as' or 'with' this squid. But perhaps that is as it should be. The technological experience reminded me of this.

In the case of the mechanical carousel, the knowledge that the creatures are not real, and larger than life, might produce a different imaginary—a lived inversion of Verne's novel, for example. And the look of these fantastical creatures—giant steel and wooden exaggerations in comparison to their living counterparts—produces a sense of awe that, in the case of the squid, but not the squib, for example, might prevent a hierarchical structure. These speculative possibilities, like Verne's 'invented worlds', may offer alternate strata of relations between humans and nonhumans—a collective Latourian 'parliament of things' with technologised representatives contributing to a greater agency for nonhumans (Latour 1993, 144).

Mediating Forces

In a world in which access to living animals is highly controlled (for the most part and especially in urban centres), the exposure to machines at Les Machines des L'île, and the ability to encounter them physically, may in fact cultivate an 'awakening' as the company's website suggests (Les Machines des L'île 2025). What this awakening is, is left open ended. The cool steampunk aesthetic returns us to a past time, to the wonder of discovery and exploration, but also to the early days of industrialisation, the early days of what has since become a climate crisis. However, Les Machines, situated on the water at the site of former shipyards, offers a chance to think about how human and nonhuman forces are mediated through technologies. They foreground narratives—such as the return to craft and seeing every step of the construction process—that have the potential to awaken all generations to alternate modes of production and to the damage that has been done to the environment and to the oceans. Perhaps the experience would be more affective if the dangers and damages the technological developments have caused could also be highlighted. Perhaps we ought not delight too much in the fantastical when actual sea creatures are coated with oil or unable to breathe due to gas leakages in their environments.

To be fair, Les Machines has been embarking on a new project, *The Heron's Tree—City in the Sky*, to specifically focus on environmental systems. Scheduled to be completed in 2027 (and undoubtedly delayed due to COVID), their press kit describes it as an 'urban park, with its own microclimate', and it will be set on the edge of the Loire River (Les Machines des L'île 2019, 12). In the spirit of Verne's invented worlds, the intention is to have flying birds on cables between the branches, and the full mechanical bestiary from within the Galerie (inchworm, heron, sloth, etc.) populating this tree. The herons will take visitors on a flight above the park. The project is being developed with the goal of 'obtaining a clean ecosystem', and it continues Les Machines' shipyard redevelopment plans around the sustainability of public art works. A port city may be a good starting place for discussions and creative imaginations of how the land meets the sea, as they propose:

> The Machines de l'île, have created a magical world using a language that Nantes residents appreciate, and which fits in with the larger history of the industrial port city open to the sea and nourished by dreams of far-off places. For a city is also built around the collective imagination of its people.
>
> *(Les Machines des L'île 2019, 20)*

While focussing on the human denizens in this press brief (which also focusses on the financing plan), from submarine life to life in flight, Les Machines is a project with the more-than-human at its heart. Looking to the sea, the river, the skies, and bringing to life mechanised inhabitants of these spaces, Les Machines imagines a world in which humans can encounter nonhuman life without destruction of these lives themselves.

To return to the opening quote, the language used in this novel of the late 1800s, full of poisoned air, disease, plague and centres of infection are all words too familiar to us in the 2020s when we are still dealing with the aftermath of a global viral outbreak and have seen the sea in flames. Fixating on Verne's (and others) invented worlds, and the technologies used to discover/conquer, collect/take, catalogue/hierarchise, has drawn attention away from the myriad lives contained in this underwater world. Recuperating the technologies to represent these creatures, to provide access to their lives without destroying their habitat, and to perhaps create better ecosystems in which nonhumans can exist, may promise a less invasive and violent future for us all.

Notes

1 Margaret Drabble (2014) has acknowledged Verne's ecological connections in an article in the New Statesman, where she also offers a wonderful description of the novel.
2 See *The Independent* (2021).
3 I am grateful to Peta Tait who pointed me to this book.

4 The park's style is often referred to as 'Steampunk' a term that suggests a hybrid style both of science fiction using anachronistic technology, and also a nineteenth-century fashion aesthetic blended with technological flair. Interestingly, 'The earliest inspirations for steampunk as a literary genre come from the works of H.G. Wells and Jules Verne. Both wrote science fiction set in their own times (the late 1800s) that featured advanced technology powered by steam, hence the "steam" in steampunk, and clockwork, including flying machines, submarines, and time machines' (Dictionary.com 2025).

5 Note the hyperbolic language that continues in a National Geographic segment on the Humboldt Squid (Nat Geo Animals 2017). It is interesting also to note that the creature Verne depicts has been translated and depicted as both a squid and/or an octopus. Both cephalopods but with differences.

6 Thanks to Kristin Tregar for pointing out these differentiations—I am still thinking through these issues.

References

Burton, Antoinette and Renisa Mawani, eds. 2020. *Animalia: An Anti-Imperial Bestiary for our Times*. Durham, NC: Duke University Press.

Dictionary.com. 2025. 'Steampunk.' https://www.dictionary.com/e/pop-culture/steampunk, accessed 1 May 2025.

Drabble, Margaret. 2014. 'Submarine Dreams: Jules Verne's Twenty Thousand Leagues under the Seas.' *The New Statesman*, 8 May. https://www.newstatesman.com/2014/04/submarine-dreams, accessed 1 May 2025.

Gerretsen, Isabelle. 2021. 'Ocean Fire Exposes Weak Regulation of Mexico's Oil and Gas Sector.' *Climate Home News*, 1 September 2022. https://www.climatechangenews.com/2021/07/07/ocean-fire-exposes-weak-regulation-mexicos-oil-gas-sector, accessed 17 January 2025.

Horn, Marek. 2021. *Yellowfin*. London: Nick Hern Books.

The Independent. 2021. 'Gulf of Mexico an Undersea Gas Pipeline Ruptured Causing the Sea to "Catch Fire".' *The Independent*, 3 July. https://www.youtube.com/watch?v=AxFqY3UTa0w, accessed 14 March 2025.

Latour, Bruno. 1993. *We Have Never Been Modern*. Cambridge, MA: Harvard University Press.

Les Machines des L'île. 2019. *Press Pack*. https://www.lesmachines-nantes.fr/wp-content/uploads/2019/04/DP-MACHINES_2019_GB_BD.pdf, accessed 23 March 2024.

Les Machines des L'île. 2025. 'The Artistic Project.' https://www.lesmachines-nantes.fr/en/the-project/the-artistic-project, accessed 1 May 2025.

Lippit, Akira Mizuta. 2000. *Electric Animal toward a Rhetoric of Wildlife*. Ann Arbor: University of Michigan Press.

Nat Geo Animals. 2017. 'Squid: The Deep Sea Devils.' https://www.youtube.com/watch?v=okBpSCqrNFA, accessed 24 April 2025.

Varney, Denise and Lara Stevens. 2022. 'The Climate Siren: Hanna Cormick's *The Mermaid*.' *TDR: The Drama Review* 66 (3), 107–18.

Vermeulen, Niki. 2013. 'From Darwin to the Census of Marine Life: Marine Biology as Big Science.' *PLoS One* 8, e54284. https://doi.org/10.1371/journal.pone.0054284.

Verne, Jules. 2020 [1870]. *Twenty Thousand Leagues under the Sea*. Ware: Wordsworth Editions.

6

BEING PREY AT THE END OF THE WORLD

Kris Verdonck in Conversation with Lara Stevens about 'Prey' (2023)

Kris Verdonck and Lara Stevens

Prologue

In 1985, Australian philosopher Val Plumwood narrowly survived an attack by a saltwater crocodile, which profoundly influenced the direction of her environmental philosophy. The following year, the film *Crocodile Dundee* was released to global, popular acclaim, cementing the association between macho Australian bushmen and dangerous crocodile predators. After her attack, Plumwood advocated for crocodile protection and her book, *The Eye of the Crocodile* (2012), is a powerful ecofeminist meditation on the fragility of human mastery over a more-than-human world that is indifferent to human narratives of identity and self-importance. Plumwood's journey through the golden eye of the crocodile shaped her influential writings in which she argues that man must overcome human chauvinism to take his place in an ecology of reciprocal, embodied exchange, decay and renewal.

In 2022, Kris Verdonck and dramaturg Kristof van Baarle visited the mountain rainforest and house where Plumwood spent most of her working life. There, they formulated the key themes of *PREY*, an exquisite music-theatre work that explores Plumwood's life and philosophy. Through a three-part performance, *PREY* interrogates the humility required for a deep understanding of the precarity of our species in a climate emergency.[1]

Stevens: Kris, you carried Val Plumwood's book *Feminism and the Mastery of Nature* around in your briefcase for years. What was it, in particular, that drew you to Plumwood's thinking?

Verdonck: First of all, I was fascinated by Plumwood's position, and, of course, her connection with nature. That is something that we humans have

DOI: 10.4324/9781003584445-9

completely lost somehow. We really do not know what nature is, especially in Europe. We have no clue what a jungle, or a predator, even means. We have lost that connection. Second, I was drawn to her attack on Plato's duality—her argument that the moment we split the relation between mind and body, we have no responsibility anymore towards the environment because we no longer have a profound connection. The story of the crocodile attacking Plumwood is guts and gore—very grim—but it is a story of beauty in that the moment the teeth went into her body, she suddenly felt profoundly connected with nature. In that brush with death, she formulates the idea of the human as being prey, part of the food system, within an ecosystem. Morality then comes under pressure in her thinking.

Stevens: Indeed and Plumwood begins *The Eye of the Crocodile* by recounting an Indigenous 'Dreaming' or origin story from the Kakadu region in the Northern Territory, where her near-death experience occurred. The story emphasises the intensity of water, stone, wind, storms and sky in this area and their power over the feeble human body. Recognition of this vulnerability necessarily shifts human morality, as seen in Australian Indigenous ethics which highlight positionality and the human obligation towards the more-than-human world. Mary Graham (Kombu-merri), for example, describes a relational Law of Obligation, which develops out of a connection to Land and notes a 'careful management of ego' as a core tenet of this philosophy (2024: 33). Graham notes that, for Aboriginal people, the relation between people and Land becomes a template for social relations, the basis for living peacefully through connection, sharing and reciprocity that contrasts with the Western short-term survivalist ethos (2024: 34).

Verdonck: Because I am a cis man, I wouldn't dare tell Indigenous stories. In the case of Plumwood, she never forgot that she was a European, or a Western philosopher, although her thinking may have been sparked or influenced by Indigenous ideas, she was writing back to Western philosophers. I thought that gave me enough permission to dive into her story.

Stevens: Yes, and it's important for Plumwood to speak to the Western audience because they are the ones who have lost that connection for the longest time. Yet the crocodile attack occurred in Kakadu National Park, a place where it is impossible for Plumwood to avoid encounter and engagement with Indigenous cultures. In Kakadu, Indigenous rangers manage a vast nature reserve, and they know the crocodiles individually, they have names for them. They know which ones are aggressive and which ones are less likely to eat you for dinner.

Verdonck: For me, there is an impossibility in that kind of thinking because I never saw a crocodile in my life. I've been to Australia a couple of times, but I've never been in this kind of situation. In her writing, Plumwood shows the impossibility of thinking like an Indigenous person of Australia since we Westerners do not have that kind of connection with the earth. The influence of Western thinking filters right down into the making of *PREY*. In theatre, the classic Aristotelian dramaturgy, with its tragic arc is the foundation of making theatre. I needed to tell a story that could have been a tragedy but, somehow, find a way to avoid that kind of Aristotelian or Shakespearean narrative sweep in the show. As a director, I grew up with all the classics, even though I never staged them. They influenced me to the point where I could no longer decide if something was good or bad or too sentimental—I was just following the system of classical European dramatisation. I had to think my way out of everything and that felt natural to me, another impossible thinking.

Stevens: You were not staging *Romeo and Juliet* in a climate emergency. You were not looking for catharsis after the hero's death. Would that make things too easy for the audience?

Verdonck: As I was thinking about Plumwood's philosophy, I was working with Japanese Noh theatre. These plays have three distinct parts. Part one, two monks walk along a path, narrating a story to one another. In part two, a ghost who has witnessed the monks' story, retells it from the afterlife. Part three is the dream scene where the two monks are asleep and the story is re-told in a highly surrealistic mode. Here the ghost can be more subjective and expressive, conveying the story with a dream-like logic, or lack of logic. The story tends to become more embodied in dance forms and ends with sounds or music that are highly abstract.

Stevens: Since Noh provided you with a way to avoid classic Aristotelian dramaturgy, could you explain how the Noh form influenced each of the three parts of *PREY*?

Verdonck: I completely followed Noh dramaturgy. We had three generations of women. Part one was narrated by an elderly and greatly respected actress, Katelijne Damen. Part two was recounted by an opera singer, Anna Clare Hauf, who was around 40 years old. Part three was the dance sequence performed by a dancer in her early 20s, Mooni van Tichel. Like Noh theatre, *PREY* reversed time. This meant that the first 20-minute monologue narrated the story of the crocodile attack without enacting the horrors.

In Part two, Hauf sang to evoke the emotions of the event, because you can be more expressive in song. Music is a very different medium for storytelling where narrative starts to be less important and you can be much more emotionally direct.

At one point in Plumwood's writings, she describes becoming a kind of shape-shifter. She interweaves her narrative of the crocodile attack with Dreamtime creatures from Indigenous storytelling. She writes about half-mermaid-half-insect creatures and the ways they link the spiritual world to our world. They are storytellers embodied in that sense, they make connections. In the third part of *PREY*, I had the dancer, van Tichel dive slowly into a mythical world where she shifts from being something very gentle to more crocodile-like, or a mythical version of a crocodile. The music becomes much more important in this part so we found another way to tell the same story that is much more abstract and mythical.

Stevens: Speaking of moving between worlds, the place where Plumwood lived and worked (and from which she took her name), Plumwood Mountain in NSW, Australia, is often described by visitors as a mystical place that exists between worlds (Plumwood 2025). Plumwood Mountain is a well-concealed 120-hectare bush property with dry and wet rainforest ecologies and climates that includes a stunning coastal escarpment. Deep into the bush there are several basic dwellings which Plumwood built in the 1970s. Plumwood Mountain is home to endangered greater gliders, spiny crayfish, lyrebirds and many wombats, with whom Plumwood recounted her intimate experiences. Could you tell me about your stay at Plumwood Mountain? What did you experience by visiting that unique place that influenced the creation of *PREY*?

Verdonck: Plumwood's house is a very, very strange place. The architecture is a 1970s futuristic vision of living in a very wild environment. One of the things that surprised us was how non-ecological this house is, for example, you must burn enormous amounts of wood and it is only slightly warmer afterwards. Everything requires so much effort. At first chopping wood is fun but, do it for a year and you would cry. Plumwood describes how difficult it is to keep books dry in the middle of the forest. The physical experience makes you feel that you do not belong there. I really felt like an intruder in the forest. I did not feel at home in any sense.

While walking around the property, I got lost at one point while I was 100 metres from the house, which was frightening. Our bodies know how to walk around in a city, I can read the city, especially as I have lived in an urban area for a long time.

I feel that if something is going on at night, you can feel the danger coming and you kind of know when to disappear from a neighbourhood. But we have no clue how to read anything in a forest, and this makes us very vulnerable.

This feeling of vulnerability in nature informed *PREY*. It was an experience I had while travelling with my girlfriend on the east coast of Tanzania, Africa. We went scuba-diving with 20 tourists out on a boat to a small island in rough seas. There was too much wind and we couldn't see any fish. The crew on the boat were not at ease there. When we got back to the boat, the guides were doing a headcount and found that one person was missing from an Asian family, a 15-year-old boy. My girlfriend and I dived back into the water and found him. We tried to rescue him, but it was too late, he had drowned. This boy was scuba-diving wearing his jeans, it all went completely wrong. He did not belong in the sea in Tanzania in bad weather. Even with good weather, he did not belong there because he could not read anything, and neither could I. We took him back on the boat and we went back on land. And somehow you felt that, for nature, nothing happened. The sea and the fish carried on. You could feel the indifference in nature, but for all of us, it was complete devastation.

Plumwood experienced difficulty belonging in that forest—although she claimed she could live there, I do not completely believe her. She was too much of an intellectual to completely belong there. Forget the idea that humans are going to stop warming the planet at one or two degrees, we are not able to. As Plumwood said, we are not connected, so we have no clue what nature is or is capable of.

Stevens: Humans can re/habituate to living with the land over time and while I don't think it was easy for Plumwood on the mountain, I also have the sense that she was incredibly in tune with its ecologies because she was curious and open to listening and learning from it. However, as you say, humans are not designed to survive alone in nature. Indigenous cultures worked with Country and its ecosystems in groups where people had particular knowledge lines and roles and it was never an individualised relationship. The idea of Plumwood living alone in such an isolated place suggests an extraordinary mental and physical fortitude.

But let us return to the boy who tragically drowned and nature's indifference. The world went on, life went on. Plumwood wrote about the indifference of the crocodile towards her identity, emotions or selfhood as it tried to consume her. This is key to her subsequent humility and philosophical formulation of the rightful place of the human in the world and the food chain. You too have had a long-standing interest in the idea of indifference, especially when deploying machines as actors onstage that are indifferent to the human judgement of their acting—they are not reading the reviews.

And yet theatre is supposed to be a place where humans can go to exercise or expand our empathy. How does this clash between empathy and indifference, and the nihilism that comes with it, work for you in the theatrical space?

Verdonck: The experience of visiting a museum or gallery is dictated by the object that you have gone there to visit. But the strange thing about the object is that we as spectators can be touched by Rothko's painting, even though it is just paint on canvas. It is extremely complex what is going on there. If we watch with our bodies, an object can tell stories somehow, it can communicate. The world of the theatre is historically a human space. One could say that this is where we rule. If you look at the history of theatre from the Greeks on, but also all Indigenous traditions, they have a very special relationship with objects onstage. From the *deus ex machina* in Greek theatre, automaton dolls that serve tea, Pepper's ghosts onstage, the half-dead, or even the dead onstage represented by an object, is something that theatre audiences, strangely enough, really like. But this is odd because one could say that it is our space. But actually, when objects go onstage, spectators develop empathy for the machine, especially one that moves a little bit.

In the history of performance art, starting with the Dadaists through to the 1970s and 1980s, Abramović and others, were obsessed with pain—cutting yourself, exhausted performers, etc. If I make a machine 'die' onstage, which is impossible, by the way, but if I 'execute' the machine, make it explode, then people sometimes say to me: 'Oh, this is really sadistic. How is it possible that you do that?'. Well, actually I am not doing anything. Nobody is injured during this performance. It's strange that people cry when R2-D2 dies in *Star Wars*. You buy a new R2-D2 and it's okay. There you go, there is no problem. But that's not the way audiences tend to feel.

So in the whole chemistry, alchemy, of a theatre piece, we project so many feelings on those objects. Because we are in that space of the living, we also project a lot of life on them. So in that sense, I was never very indifferent. As human beings do not have so much imagination, all robots have the form of animals. If you go to a robot exhibition, it's all snakes, dogs and birds. We can't apparently imagine others. Or human beings, dolls, or cyborgs. If I put a dog onstage, or a walking robot onstage, we have more or less the same feeling due to the 'uncanny valley' phenomenon. We project life on the objects, and it can become uncanny. But if you would compare, a little mouse onstage, everybody would look at the mouse because it's so innocent and it draws all the attention, like a baby. You've got the three Bs. There are babies onstage, beasts onstage and a third one—I'm sorry but it's true I'm afraid—is breasts onstage. They steal the first ten seconds. But anyway, let's stick with the beasts and the babies. We have the same relationship with spontaneity, naivety or the non-conscious, the half-conscious, of animals onstage, or children onstage. They seem spontaneous, they have no ego. If you take von Kleist's 'On the Marionette Theatre', well, there you have it. The big problem with actors is that they act.

Stevens: What does this mean for the use of Artificial Intelligence for future Verdonck/A Two Dogs Company performances?

Verdonck: AI is as stupid as other machines. The narrative now is that it is so intelligent, and we project so much onto it, as if it understands language. It analyses hundreds and thousands of texts, and this makes it seem as if it is 'thinking' or 'understanding'. But I am currently making a work that uses AI.

Stevens: In a world where we are increasingly allowing AI to do our 'thinking' for us, what do you think that you add to Plumwood's philosophy through embodied performance? What does theatre bring into the frame?

Verdonck: I see *PREY* as an introduction to Plumwood's texts more than anything else. I want to invite young people to get to know her, how beautiful her texts are and to show them the value of her critique of patriarchy and Western philosophy, and the systems upon which European culture is built.

Note

1 Further information about *PREY* can be accessed via the artist's website. https://www.atwodogscompany.org/en/projects/prey/.

References

Graham, Mary. 2024. 'The Relationist Ethos and the Law of Obligation.' In Anne Poelina, Donna Bagnall, Mary Graham, Ross Timmulbar Williams, Tyson Yunkaporta, Chels Marshall, Shola Anthony Diopm, Nadeem Samnakay, Michelle Maloney, Michael Davis, *Declaration of Peace for Indigenous Australians and Nature: A Legal Pluralist Approach to First Laws and Earth Laws*. Singapore: Springer, 31-41.

Plumwood. 2025. Website. https://www.plumwood.org, accessed 30 January 2025.

Plumwood, Val. 2012. *The Eye of the Crocodile*. Edited by Lorraine Shannon. Canberra: ANU E Press.

7

ECOSPHERIC AFFECT IN *SOLO POUR OCTOPUS*

Sarah Lucie

In response to the climate crisis, Una Chaudhuri and the Climate Lens network have issued a call for an 'ecospheric' and 'dis-anthropocentric performance'. The theatre practice these performance-makers and educators imagine goes beyond representing the effects of climate change in the world or addressing the feelings of anxiety that emerge as a symptom of climate change. Instead, Climate Lens asks for 'a dis-anthropocentric, ecospheric ideology and art practice [that] seeks to partner with the more-than-human world and [does] so in the full acknowledgment that our species has changed the planet in profound ways' (Chaudhuri with Climate Lens 2023, 41). Underlying this call is a commitment to undo anthropocentrism, including its hierarchical systems and 'fictions of authoritative intellectual mastery' (41) of the world. The project warns against mapping out a specific path forward—a planned utopia can be manipulative, after all, and a problem-solving sense of progress upholds an anthropocentric sense of mastery. Rather, as Chantal Bilodeau reminds us, 'the climate crisis has been called "a crisis of imagination"' (Bilodeau n.d.), and Chaudhuri believes theatre practice is suited for affecting this broader way of thinking and seeing. Thus, the group aims to collect and develop strategies for undoing anthropocentric fictions, in this space that is so inherently anthropocentric: the theatre. The Climate Lens playbook is organised as strategies towards creating work that is meaningful in the face of the climate crisis, and I begin with their call to arms to share the lens from which I imagine theatre's potential and analyse the work before me. Beyond theatre creation, their framework also helps theatremakers and scholars to value different elements of theatre, to consider alternate plot structures, settings, characters and atmospheres, and to think creatively about a relevant theatre for the future. The full playbook offers 16 directives, but I

DOI: 10.4324/9781003584445-10

highlight a few here that have captured my imagination and guide my ensuing analysis of *Solo Pour Octopus*.

1 Practice literalism. End the tradition of turning everything into a symbol for human life.
2 Occupy science. Befriend facts and factoids. Enrich theatre with the bristly nomenclatures of the natural sciences.
   [...]
11 De-sentimentalize 'Nature'. Keep the awe, lose the 'Awww!!!' Forge new affective pathways to the non-human — beyond sadness, guilt and fear. Invite in humour, anger, joy, irony, sarcasm....
12 Stand alongside our fellow species like a breathing exercise, to open up space in our cells for epistemologies of the biosphere that our bodies currently don't hold, or ones we need to re-ignite. Physicalize awe.
16 Create theatres of species life; fill the stage with the Earth (Chaudhuri 2023, 42–43).

I find myself returning to the directive to forge 'new affective pathways', as this practice is essential in the larger project of finding alternative and multiple ways of knowing, especially ways of knowing that chip away at the Humanistic focus on reason and progress. For instance, Mel Chen suggests that affect functions as communication from non-human matter commonly understood as inanimate, an alternative to linguistic communication and the humanistic hierarchy that comes with it (Chen 2012). It is easy to imagine how a germ, a radio wave or a flower have the capacity to affect their environment, create change and be affected in return, and even how these might invite feelings or responses from human bodies. Creating theatre while attuned to non-human affect deprioritises language and narrative but draws attention to the feelings and sensations created in the performance-audience interaction. Working with affect as a dramaturgical tool, then, draws focus to non-human agency and opens a direct line of communication between a human audience and non-human actors regardless of reason or language. Furthermore, to forge 'new' affective pathways also involves a sense of dis-orientation. If affect is an essential part of the way humans relate to the world, build habits and engender identity and sense of self, then to dis-orient, as Sara Ahmed has suggested, can ignite new formations of relation. Indeed, 'moments of disorientation are vital' in that they are alive, actively re-mattering the world (Ahmed 2006, 157). Sensory experiences have a direct connection to perception, but also to the ways we engage with ourselves and relate to the world. These affective pathways may invite, or establish and practice, alternative senses of the self, revealing ways of being alternative to the Humanist myth of the wholistic 'I' governed by reason.

With the Climate Lens's ecospheric imperative in mind, I turn to Stefan Kaegi's production of *Solo Pour Octopus*, which premiered on 13 May 2021 and was performed for a live audience and livestreamed before being turned into a film. In the ensuing chapter, I will analyse the ways this production engages with many ecospheric theatre goals, including that of 'forging new affective pathways'. I am particularly interested in how these affective pathways are built even through its livestreamed and filmed iterations, when an audience is not sharing a physical space. The affective nature of the production even through a screen suggests, maybe paradoxically, that highlighting our humanness and anthropocentric practices through a pointedly constructed, metatheatrical event may be a useful tool in dis-anthropocentric performance, adding another tool to the climate theatre playbook.

The *Solo Pour Octopus* title suggests that the performance is a 'solo', and it does indeed centre a living octopus onstage, as the 'protagonist' if you will. The performance begins with lights coming up on an aquarium, lit so that we, the audience, can observe the octopus from our position in the seats or through the screen. But, of course, the octopus is not alone. The design of the production features a live feed video of the octopus projected behind the aquarium, as well as improvised music that responds to the octopus's movement and an audio recording of multiple texts offering facts about octopuses and other reflective themes. But there is even more direct contact with the octopus beyond these atmospheric elements. First, a single light spot appears in the aquarium, which the octopus visibly responds to by moving away from the light's boundaries. From there, the interaction grows to a larger sphere of light. And finally, a human performer (Nathalie Küttel) joins the stage. Her proximity to the octopus steadily builds as she moves close to the aquarium, touches the outside, touches the water and, eventually, puts her whole arm—then her face—directly into the water. The solo is not a solo, but an intimate, touch-based call and response between multiple human and non-human materialities.

As my description already suggests, the production engages with a few of the items in the climate lens playbook. Most obviously, the very fact that a literal octopus appears onstage, not as a metaphor but as an octopus, reorients the sense of protagonist. Some of the texts in the voice recording also explicitly remind the viewer to resist anthropocentric understandings of the octopus and help to 'befriend facts' (number two of the playbook). For instance, one section describes the octopus's nervous system and the way their arms engage with the world. Rather than one central brain—one 'I'—the distinct nervous systems in the arms respond directly to their input. The octopus displays a different kind of responsiveness to the world. Another text suggests that it is unfair to try to understand octopus intelligence in a way that is parallel or comparative to humans. And, embodying number 12 of the playbook, Küttel literally 'Stand[s] alongside our fellow species' as a collaborator onstage, as they take each other in and respond through what could be considered their own kind 'breathing exercise'. Kaegi shared that the octopuses 'define much of the action, interaction

FIGURE 7.1 Nathalie Küttel interacts with the octopus in *Solo Pour Octopus*. Image courtesy of Stefan Kaegi.

and dramaturgy of this show and it is new every night' (Kaegi et al. 2021), and a part of the rehearsal process involved Küttel training to listen to the octopus during their offstage work. Küttel and the octopus developed their relationship over the long-term to build trust towards an actual partnership (Figure 7.1). The team actually rescued two octopuses, both from a food market in Sète, France. Often fishermen hunt with harpoons or through methods that kill the octopus on contact, but in Sète, a fisherman fishes with pots and the octopuses remain unharmed. When the octopuses first arrived in Gimel—the aquarium/installation set up by Kaegi, Küttel and Judith Zagury—Küttel spent a few nights sleeping in a tent next to the aquarium, and in the early weeks, both Küttel and Zagury were in daily contact with both octopuses to get to know them (*Temple du présent* team 2021, 40). The octopuses worked with the team for three months before the technical elements were introduced onstage.

While the relationship between the octopus and human performer captured my attention, the most overwhelmingly ecospheric element of this production, for me, was the affective experience. Dramaturgically, this incremental broadening of focus and integration of new elements helps to direct audience attention, making each individual moment of increasingly intimate points of contact between the octopus and the human monumental. I watched with a sense of anticipation and a heightened focus tuned into the exact moments of contact. In my experience, this heightened awareness of the actor's physical touch felt intimate and sensorial even through the streaming platform, as if the human performer was an avatar for myself—moving forward with curiosity and wonder. I lived vicariously through her touch, as we watch Küttel's fingers move through the water. Similarly, when the octopus wraps one arm around her fingers, I watched with rapt attention that I felt in my whole body. Of course ocular senses were engaged, but goosebumps prickled and my stomach seemed to clench. My sensory perception was working overtime.

My multisensory experience is not revelatory—in fact, the synchronic sensual response is often discussed in cinema studies. Luis Rocha Antunes suggests 'there is no such thing as a purely visual—or purely audiovisual—experience of film. What we *see*, or what we call *seeing*, is multisensory in its nature' (Antunes 2016, 3). He argues that what we experience by watching a performance doesn't derive solely from interpreting the information through our personal filters of memory or imagination, but through 'actual perceptions that are endogenously generated under the influence of actual sensory information from the external context' (39). This sensory perception occurs through affect, created by the theatrical or cinematic elements; character mediation, or what the bodies on view do and experience; and then this personal interpretation. Following this theory, the mediated theatre form and the physical distance between the performance and the audience do not preclude an audience body's reception of sensorial information.

That this ecospheric performance is able to affectively and sensorially touch its distant audiences through the screen, and that theory from cinema studies is so relevant, prompts an interesting tension regarding the nature of ecospheric performance—namely, the mediated nature of the theatrical setting. *Solo Pour Octopus* reminds me of Joseph Wright of Derby's painting 'An Experiment on a Bird in the Air Pump', which depicts a moment in the Enlightenment where scientists attempt to discover 'truth' through their scientific experiments only by taking 'nature' out of context and sealing it off in a man-made vacuum (Figure 7.2). The similarity between the two images is striking, as the artists prompt the audience to understand the animal—bird and octopus—through a completely manufactured and controlled setting, while the animal is distantly removed from its habitat.

This tension between the ideal of an untouched nature and the reality of a cultural frame is clear in Stefan Kaegi's writing and speaking about the project. He reveals a wish for, or perhaps naïve sense of, a 'pure' and natural animal. He comments on the 'non-artificiality of what an animal does on stage', and its ability to '[open] a window into reality' (Kaegi et al. 2021). This sentiment is a frequent refrain regarding animals onstage. As Rachel Price points out, aesthetic theory has held up the animal as an 'honest' being without a theatrical, performative self (2006). The thinking goes that an animal doesn't experience a split subjectivity, or the separation of mind and body that produces an individual, self-aware 'I' in humans, so they are incapable of lying by performing any

FIGURE 7.2 Joseph Wright of Derby's painting 'An Experiment on a Bird in the Air Pump' (1768). National Gallery, London. The painting depicts a bird at the centre, in a vacuum. Public domain.

separate self onstage. However, this thinking builds on the idea of a mind/body split that is frequently challenged by contemporary philosophies. Materialism, on the other hand, suggests that the mind is a product of the physical form and emphasises the performative agency of matter to effect and affect their environments. A materialist perspective finds more similarities between human and animal being—both sensing through bodies that are embedded in and responsive to environment. This materialist view, therefore, suggests that nothing is untouched by its environment, which begins to trouble the possibility of an animal's supposedly authentic nature onstage.

The dramaturgy and design of the performance further disrupts Kaegi's idea of a pure animal, by emphasising the constructed aquarium in a constructed stage. The stage lights are visible, as are the filming mechanisms that bring the closeup of the octopus to the backdrop of a screen. Furthermore, what we see onstage is the result of the relationships that developed over time between two singular individuals. Therefore, the theatricality and the artifice onstage reflect the actual artifice—the constructedness of our knowledge more generally. As Alain Kaufman, a specialist of biology and sociology, points out in conversation with Kaegi, scientific research may aim to avoid disturbing the object of study, but in all behavioural studies, the skills are dependent on the quality of the affective relationship of the animal with the researcher, and the worlds of both the animal and trainer are enriched by the collaboration (Kaufman and Kaegi 2021, 21–26). It is *this* reality that is staged—not some truth of octopusness, but our constructed and relational knowledge and experience of the world.

Octopuses feature in a number of other recent artworks, which similarly make false claims at revealing untouched nature. For instance, the 2020 Netflix documentary entitled *My Octopus Teacher*, which won the Academy Award for best documentary feature in 2021 among numerous other notable awards, follows Craig Foster as he trains as a free diver in South Africa. He meets a young octopus and quickly becomes enamoured with it, visiting her regularly enough that the octopus becomes more comfortable with his presence. The documentary frames their relationship as one of teacher and mentor—where the octopus teaches Foster about nature and the fragility of life. In practice, the story is more about the relationship than the octopus itself, and the documentary never honestly addresses the fact that the diver is obsessed with the octopus and invades its space, unwilling to leave it alone. In other words, while the documentary claims that the octopus 'shares the mysteries of her world', another summary might describe it as 'a man uses an octopus for his personal interests and needs'. Some reviewers pick up on this power dynamic. Julia Furmanek's review in *The Daily Free Press* begins with the summary: 'the story of a magnificent creature … and the egotistical man who won't leave her the hell alone' (2021). The documentary's footage grants the audience access to a free diver's ocean views and an octopus outside an aquarium, creating an enticing and mesmerising experience. However, it anthropomorphises the octopus as Foster's feminine companion without interrogating Foster's obsession and the reality of the human influence in the ocean setting—or the human sense of privilege and entitlement under it all.

These two projects' choice to centre the octopus—whether onstage in an aquarium or while freediving—out of all the other potential subjects points to a broader fascination with and potential of this particular animal.[1] Octopuses are known to be extremely intelligent, often curious and inventive, which make them especially dynamic animal protagonists. But also, the materiality of the octopus's body, with its eight arms covered with suckers, makes the octopus an especially affective protagonist that induces physical sensations in many viewers. In fact, some may experience anxiety when faced with an octopus's arms, or an image of it, due to trypophobia—an aversion to clusters of small holes. According to Geoff Cole from the Center for Brain Science, 'Approximately 10–18% of the adult population experience some form of anxiety when viewing clusters of small holes' (2024), while the negative feeling Cole focusses on most is 'disgust'.[2] As someone who advocates for empathy and respect towards non-human animals, the notion of purposely igniting disgust, which often leads to distance, against a non-human performer seems antithetical. However, Sianne

Ngai argues that the ugly feeling of disgust is not all bad. Rather, disgust gives power to the object. In other words, disgust may be better than 'an object perceived as so unthreatening in its inferiority as to be barely perceptible at all' (349). Disgust puts the object at the centre of one's perception so that one cannot remain indifferent. Though Ngai admits 'disgust does not so much solve the dilemma of social powerlessness as diagnose it powerfully' (353). That both of these octopus artworks stand to ignite and harness this potential disgust is perhaps an ecocritical project itself, doing away with indifference.

However, disgust is only one potential affective response. Many reviews and audience feedback of *My Octopus Teacher* picked up on Foster's obsession with the octopus, which some categorise as erotic (Elle Hunt ruminates on how and why the interest in Foster's 'erotic relationship' went viral in her review [2020]). Like *Solo Pour Octopus*, *My Octopus Teacher* features closeups of the octopus reaching out and touching human flesh, and in the documentary this moves to full body contact when the octopus sits on Foster's chest. *Solo Pour Octopus* dramaturgically builds desire, through the intimate focus on the octopus's movement with the incremental addition of Küttel's touch into the shared space. By building anticipation, the first touch between Küttel and the octopus fulfils a desire that had just been developed. Desire is the reciprocal affect of disgust (see Ngai 2007). Whereas disgust 'strengthens and polices' (335) the subject-object boundary and suggests a 'rejection or *exclusion* of its object' (22) that requires the subject to look away, desire, on the other hand, suggests intimacy and a wish to close the distance, to remove a subject-object boundary. In this way, perhaps both examples of the performing octopus create a reorientation of the senses, opening pathways of affective, sensorial communication with the octopus that render the octopus able to be seen. They both showcase an intimate relationship between an octopus and a human collaborator, with a dramaturgical affect that invites the audience into experiencing the affect for themselves.

What then creates the ecospheric affect of *Solo Pour Octopus* as opposed to *My Octopus Teacher*? It's not only the difference in tone, created through *Solo Pour Octopus*'s voiceover that shares facts emphasising the human-octopus differences and asks the audience not to anthropomorphise the octopus, nor is it only the difference in setting, where we meet an octopus held in an aquarium versus a recording of the octopus in its habitat, but it is also the self-aware theatricality of the event. At one point, the camera that has been filming the octopus turns downstage, and we see through the aquarium's glass to look at the human bodies in the audience. With this simple move, the act of human perception becomes an additional theme. Whichever affective response—disgust, desire or something else—has been kindled in the audience, it only occurs in relation *between* the human and animal actors. In other words, the audience are implicated in the meaning making and creation of knowledge in this event. Yes, *Solo Pour Octopus* is about an octopus, but even more, it's about the constructed nature of each individual's knowing of the world around them. And ecospherically, the way of knowing in this production pointedly involves affective sensation, where humans receive direct information from the octopus through their bodies as well as from the text and dramaturgy.

Much ecological and ecospheric theatre looks to site-specific performance and environmental or immersive staging to access the different pathways to audience imagination and affective experiences. However, this kind of performance isn't always possible, either due to accessibility concerns, strict safety measures to promote health in a pandemic-prone world, or simply the logistical challenges and audience expectations in university or regional theatre settings that make more traditional theatre with a familiar production timeline preferred. It is essential to remember, then, that theatre staged in a black-box or proscenium—a cultural vacuum set apart from 'nature'—or mediated through technology on a streaming platform still has ecospheric potential. In fact, in these mediated settings, the explicit embrace of a particular, situated human perspective is an effective tool to reveal anthropocentric assumptions. Chaudhuri argues that 'to undo anthropocentrism we must refuse its hierarchical and centralised systems, its disembodied and alienated epistemologies, and its fictions of authoritative intellectual mastery from a synoptic God's-eye view' (Chaudhuri 2023). When I first read this statement, I couldn't see how a traditional theatre set-up, the synoptic, hierarchical view of a proscenium stage, could begin to approach the ecospheric tenets. After the experience of *Solo Pour Octopus*, ecospheric performance seems more adaptable than I first thought, offering affective reorientation in a proscenium, even through a screen across thousands of miles.

Notes

1 Another popular example is the novel *Remarkably Bright Creatures* by Shelby Van Pelt (2022), which follows Tova as she develops a relationship with Marcellus, a giant pacific octopus living at the aquarium where she works the night shift. As of this writing, a film adaptation is in development by Netflix.

2 This disgust may also be a survival tool. Eating a live octopus, or an uncooked and freshly detached arm, poses a choking hazard as the suckers may attach to the inside of the throat. See *The Korea Herald* (2023).

References

Ahmed, Sara. 2006. *Queer Phenomenology: Orientations, Objects, Others*. Durham, NC: Duke University Press.

Antunes, Luis Rocha. 2016. *The Multisensory Film Experience: A Cognitive Model of Experiential Film Aesthetics*. London: Intellect.

Bilodeau, Chantal. n.d. 'Theatre in the Age of Climate Change.' *Howlround Theatre Commons*. https://howlround.com/series/theatre-age-climate-change, accessed 1 March 2025.

Chaudhuri, Una, with members of Climate Lens. 2023. 'Dis-Anthropocentric Performance: The Climate Lens Playbook.' *TDR* 67 (1), 41–43. doi.org/10.1017/S1054204322000740
Chen, Mel Y. 2012. *Animacies: Biopolitics, Racial Mattering, and Queer Affect*. Durham, NC: Duke University Press.
Cole, Geoff G. 2024. 'Is Trypophobia Real?' *BJPsych Open* 10 (2), 1–4. doi.org/10.1192/bjo.2023.621
Furmanek, Julia. 2021. 'Review: "My Octopus Teacher" Is Not for the Faint of Heart.' *The Daily Free Press*, 12 April. https://dailyfreepress.com/2021/04/12/review-my-octopus-teacher-is-not-for-the-faint-of-heart, accessed 1 March 2025.
Hunt, Elle. 2020. 'An Octopus "Love Story" on Netflix Has Caused Thoughts to Run Wild. Why?' *The Guardian*, 24 September. https://www.theguardian.com/commentisfree/2020/sep/24/octopus-love-story-netflix, accessed 1 March 2025.
Kaegi, Stefan, Judith Zagury, Nathalie Küttel, and Kerstin Schroth. 2021. 'Conversation between Stefan Kaegi, Judith Zagury, Nathalie Küttel and Kerstin Schroth.' *Moving in November*, 12 August. https://movinginnovember.fi/conversation-between-stefan-kaegi-judith-zagury-nathalie-kuttel-and-kerstin-schroth, accessed 1 March 2025.
Kaufman, Alain and Stefan Kaegi. 2021. 'We Are Split When It Comes to Animals.' In Stefan Kaegi, Judith Zagury and Nathalie Küttel, eds, *Temple du présent – Solo for octopus*. Lausanne: Théâtre Vidy-Lausanne, 21–26.
The Korea Herald. 2023. 'Korean Delicacy of Live Octopus Leads to Choking Death of 82-Year-Old.' *The Korea Herald*, 24 October. www.koreaherald.com/article/3239820, accessed 1 March 2025.
Ngai, Sianne. 2007. *Ugly Feelings*. Cambridge, MA: Harvard University Press.
Price, Rachel. 2006. 'Animal, Magnetism, Theatricality in Ibsen's *The Wild Duck*.' *New Literary History* 37 (4), 797–816.
Temple du présent team. 2021. 'Making Theatre with an Octopus.' In Stefan Kaegi, Judith Zagury and Nathalie Küttel, eds, *Temple du présent – Solo for octopus*. Lausanne: Théâtre Vidy-Lausanne, 34–43.
Van Pelt, Shelby. 2022. *Remarkably Bright Creatures*. New York: HarperCollins.

8
PERFORMING LUNAR CLIMATES

Felipe Cervera

The first plant on the Moon was a cotton plant that lived on the satellite's far side onboard the Chinese rover Chang'e 4 (Fox 2019). It first grew on 15 January 2019 and died only a few days later. It sprang up in a world that was not its own—it was born worldless. Its given environment was already its death sentence, designed to test terrestrial life's resilience in a climate not its own. The plant was part of an experiment called Biological Experiment Payload, which included six components: plants to produce oxygen and nutrients through photosynthesis, fruit flies to consume those nutrients and yeast to decompose the waste from the flies and dead plants to create a self-sustaining microsystem. This was indeed an advanced effort to test terraforming methods on the Moon. Originally intended to last as long as a hundred days, the experiment was terminated nine days after the plant sprouted, when external temperatures dropped to –190°C, and the plant finally died. The fruit flies never hatched. The plant remains the only lunar native we know of—the only example of extraterrestrial life on the Moon.

The first plant on the Moon died tragically, having no other life around it to communicate with, no other performer with whom to enact a planet of their own, no meaningful relationship to make other than with its finitude. It was sent to the Moon to die alone, as Agambenean 'bare life' sacrificed in pursuing the human Heliopolis and its spectacle of science and mortality. The plant is part of a long series of experiments, both in fiction and reality, performed to spring life outside of terrestrial climates. Dogs and monkeys rocketing through the Earth, plants and flies orbiting it on board space stations, plants springing on the Moon: terrestrial life seems to perform as a force to be reckoned with by the rest of the solar system. These experiments are, in principle, performed to advance life.

DOI: 10.4324/9781003584445-11

But the opposite may, therefore, be true, and they are performed to test death. The first plant on the Moon was never intended to survive. It grew not to live but to die. The guiding question of the experiment can thus be phrased: how long will this ecosystem live? Or, put differently, how long before it dies? How far out can they, the humans who can fly, eventually claim?

The binary between life and death is an initially viable framework through which to understand how extraterrestrial biopolitics/necropolitics often determine, on the one hand, the imaginaries of lunar climates as deathly scenarios and the cultural frameworks through which space colonisation is narrativised, on the other hand. To be sure, terrestrial life cannot survive by itself 'out there'. Extraterrestrial space is deadly for us and the majority of other terrestrial non-humans. But equally as mortal is the terrestrial atmosphere—it is just that life as we know it survives longer 'here' than 'there'. And the allegorical hermeneutics of lunar landings are diverse, so it would be inaccurate to say that all lunar performances are necropolitical. Rather than merely sanctioning a correspondence between life and death as equivalent to good and evil, the premise here is that examining how the relationship between climate and death is depicted in space, both in science and fiction, may serve as a powerful entry point for articulating a more contemporary extraterrestrial aesthetics for the performing arts—one with enough political significance to contribute to the discussions surrounding an increasingly necessary celestial and space (post)humanities.

The first plant on the Moon was the protagonist in a climatological rehearsal, a first sketch of what climate change might mean for and be like on the Moon. In this sense, the first plant on the moon was the product of a lunar staging—the construction of a scenario meant to reinstate and actualise a 'meaning making paradigm' to structure the 'social environments, behaviours, and potential outcomes' (Taylor 2003, 26) of an eventual, though not fictitious, effort to terraform parts of the Moon. In parallel to its scientific purpose, the plant was also performing another role, one that is perhaps more clearly envisioned as symbolic or discursive, but that is nevertheless equally powerful to communicate the advancement of a human-centric Heliopolis beyond planet Earth. The staging here, therefore, works in a double bind: at once, performing a scientific experiment in a carefully calculated series of steps aiming for a specific outcome and, at the same time, the actualisation of an imaginary in which extraterrestrial life becomes viable as an extension and derivation of the terrestrial. And in both ways, the plant, despite its death, stands triumphally at its bow: It gave the scientists the data they sought, and, on the other hand, if a cotton plant may survive on a celestial body other than Earth, so might humans. An extraterrestrial or post-terrestrial human-friendly climate is possible. The first plant on the Moon was a stand-in for the lead actor, an eventual lunar resident, who, for now, sat patiently on Earth watching the plant die.

Here, extraterrestrial space does not refer to where 'aliens' come from. It refers to the social space, inhabited mainly by non-humans, that now expands to the solar system and beyond. It is important not to consider the musings above as a mere speculation or a rhetorical device. The main challenge to study outer space exploration critically and for various activist and scholarly purposes is to stop considering items such as 'lunar resident' or 'terraforming the Moon' as phenomena existing in a forever yet-to-be—a fiction, a hyperbole, a dream. At the time of writing, the corporatisation and privatisation of space endeavours means that the 'space industry' is currently valued at over US$600 billion and is projected to triple that and reach a worth of $1.8 trillion within the next ten years (Acket-Goemaere et al. 2024). The current richest man on planet Earth, Elon Musk, owns, via his company Starlink, a network of low-earth orbit satellites that currently has over 7,000 satellites in orbit and plans to launch 35,000 more in the coming years. Space debris is increasingly becoming an ecological and scientific problem as orbital pollution reaches dangerous density (Clormann and Klimburg-Witjes 2022). As dark skies disappear, scientists face greater difficulty in achieving astronomical observations without obstacles. Lastly, space law is going through a critical moment in light of the Artemis Accords and the revitalisation of the question of whether or not it is legal and ethical to mine the Moon and other celestial bodies. It might take a few decades, perhaps even a century, before we see a full-blown lunar industry, but it would be a gross mistake to stop short of studying outer space as a real and material dimension of planetary politics *today* and focus instead on its fictional imaginaries alone. Now more than ever, access to the sky is at stake. How we perform that access matters—off- or onstage, is the same.

Outer space has parallel lives as science fiction and as a social reality. And perhaps like no other domain of scientific and political projection, fictional space exploration has set a historical precedent for the social production of outer space ever since Jules Verne's fiction inspired the work of early rocket science at the hands of Konstantin Tsiolkovsky, who, in turn, and inspired as well by Soviet cosmism, saw in space colonialism the natural fate of humankind. He saw a future without death (Scharmen 2021). However, it is equally important to understand that the connection between the real and the fictional, the political and the aesthetic, in matters concerning outer space, is not just a consequence of the advent of space flight during the twentieth century. Space exploration is born from the fictional and the ritual realms of cognition before it takes root in the technoscientific. When we speak of lunar stagings, therefore, we may not only include technoscientific scenarios, such as the first plant on the Moon, but also the many cases in which the Moon is appealed to as a part of the terrestrial experience, either ritualistically or dramaturgically. Lunar performances are central to the development of human cultures because the Moon is central to the terrestrial experience. And it will continue to be so as post-climate change culture and subjectivities shape up in the twenty-first century and beyond. How we perform the Moon matters as well. It is, after all, a key symbiote in our solar ecology.

Oceanic tides result from the gravitational relationship between the Earth and its natural satellite. Entire species that live in the intertidal zones, such as sea turtles, are so tuned into the tidal rhythm that they would probably go extinct should the Moon disappear. Life on Earth would be very different.

Fred Scharmen explains how

> the creation of new environments is always the creation of a new subject [and so] when climate change on Earth is reframed as a question about which subjectivities the planet's environment is for, we can see [that] if the environment is designed, then the population is designed.
>
> *(2017, 185, 192)*

From this point of view, the history of outer space presents itself as a history of climate change. This can be understood in two senses. Firstly, the development of space science and technology has been pivotal for advancing climate science and, therefore, instrumental to our current knowledge about climate change on Earth. In a second sense, however, exploring outer space requires creating and sustaining artificial environments. In this second sense, we can describe lunar performances where terrestrial life is involved, fictional or otherwise, as the *staging* of a climate change scenario. The first plant on the Moon was indeed a post-climate change test subject. It sprang in a world where it could not make sense of its environment. It was sent there to test the meaning of life indeed. It was displaced before emerging—*la vie avant la vie*, a genuinely *radical* avant-garde, bare life by climatological design.

Rehearsing Climate Change and Extraterrestrial Biopolitics in Performance

What is there to learn about climate change, its imaginaries and technoscientific scenarios if we study the performances of lunar climates? To the extent that climates produce subjectivities, studying how lunar climates have been and are staged may shed light on how post-climate change subjectivities are shaped on Earth or elsewhere *now*. Consequently, this perspective may foster the sensitivity to identify the type of subalterity produced thereafter. A performance history of lunar climates can contribute in this way to the conversations about race, colonialism, climate violence and displacement that we need now on Earth. It is time we let go of the Moon as a fantasy, a sci-fi trope. The Moon and its climate are at stake today because Earth is at stake, too. The sociotechnical sphere of our species is already extraterrestrial—beyond Earth—and we are no longer an 'inner-space' species. Humans sprang life on the Moon in 2019. A debate on extraterrestrial biopolitics is urgent.

A look at George Méliès's 1902 movie *A Trip to the Moon* is helpful to the argument. Produced in 1902, the film is considered the first science fiction movie. Méliès's film is important in that regard because it inaugurates, *avant la lettre*, a space aesthetic that departs from the theatrical and enters the cinematographic. This is an important detail to consider, even when it pertains to a conversation longer than the one afforded in this chapter, because that departure from the theatrical meant live performance became foreign to the more extensive articulation of space-age/astronautical aesthetics in the latter years of the twentieth century and first quarter of the twenty-first. Indeed, Méliès's film is, as film historian Thierry Lefebvre puts it, a 'composite film', that drew inspiration not only from Verne's lunar travel novels, as it is widely known, but also from live-action performances from varying genres and origins. Notably, Lefebvre traces these influences to Jacques Offenbach's operatic version of the tale, titled *Le Voyage Dans la Lune* that was first produced in Paris in 1875, a performance lecture titled *A Trip to the Moon* by an author he does not name and that was performed in Berlin in 1887, and lastly to the ride *A Trip to the Moon*, which was the main attraction at the 1901 Pan-American Exposition in Buffalo NY, USA (Lefebvre 2011). Indeed, Méliès's version of the story was not the first one, but it wasn't either the consequence of an imagination rooted in fiction alone. Instead, it was part of the more extensive cultural and technoscientific impulse that initially included live performance, and that would eventually, but not too long after, give rise to astronautics. Only a couple of months after Méliès first showed his film, Konstantin Tsiolkovsky, widely regarded as the founding figure of astronautics, published his germinal work, *Research into Interplanetary Space by Means of Rocket Power* (1903).

The historiography of lunar travel performances noted above indicates that performing trips to the moon is a phenomenon rooted across different media and shaped by different epistemes. This is the interdisciplinarity that directly shapes the history of astronautics. When Méliès produced the movie, space exploration was not yet a material reality. But that does not mean that the possibilities of space flight, and an eventual lunar landing, were still exclusively a product of fictional imaginaries. These dreams were becoming plans. It is, therefore, important, and emphasising once again a point mentioned above, that we approach lunar performance not only as a metaphor or hyperbole but also as part of a larger history of rehearsing planetary differences and the subjectivities that difference may bring about. In these examples, we find a proto-imagination of planetary diversity as climatological difference. From this perspective, an analysis of Méliès's *Trip* will shed light on the formation of a biopolitical strata in producing extraterrestrial life and, therefore, on the anthropocentric biopolitics/necropolitics of space and planetary exploration.

The plot in Méliès's film is well known: a group of scientists (comically dressed as magicians) decide to use a giant cannon to propel themselves to the

Moon. Once there, they find themselves amid rocky terrain, and seeing that night is upon them, they decide to rest. A comet passes by, the stars fade in, and shortly after, a snowstorm falls. Awoken abruptly by the snow, the scientists find shelter in an underground cave. There, they are met by a Selenite—a fictional lunar native—who is awkwardly portrayed as an insect-like humanoid and who jumps out from a giant mushroom. Scared and frantic, one of the scientists attacks the Selenite and kills them. A squad of Selenites appears next. They capture the group of human scientists and take them to see their king. Barely seconds after being introduced to the king, the human scientists release themselves and start attacking the Selenites, including the king, whom they kill. The humans escape, find their bullet capsule and manage to 'drop down' back to Earth. A Selenite finds them at the last second, latches onto the bullet capsule and travels back to Earth with them. Back on Earth, the scientists are received as heroes by a crowd who also becomes frenetic once the Selenite, now chained, is displayed as a bounty, in what is a comedic reference to the practice of displaying Indigenous peoples at colonial world fairs. Already, a century before the cotton plant on the Moon, extraterrestrial life is dealt by Méliès with a hand of death cards.

Méliès's *A Trip to the Moon* is a violent staging of the Moon. It stages interspecies violence that displaces lunar life and disposes of it as expendable in the name of human science. The film thus presupposes an early necropolitical disposition in astronautics to kill non-human life at the slightest provocation. This follows what Treviño has explained as the colonial cosmic order of coloniality, which

> separates nature from humanity, enforces a hierarchy of humans, and renders nature, and those denied person-hood, as fully exploitable. Under this cosmic order, all relations and ways of being and knowing are reduced to relations of exploitation; the heavens are used to legitimise this.
>
> (Treviño 2023, 226)

Méliès's lunar staging performs a planetary imagination that visually connects modern science and technology with the kind of male-white Eurocentric subjectivity Treviño focusses on as she develops her critique of extraterrestrial coloniality. Centrally for the argument here, however, the interspecies encounter in *A Trip* does not happen in a climatological vaccum, for the Moon is staged as having a climate very similar to Earth's extreme environments. In this sense, the connection between the film's colonial, racial or interspecies violence has not been made, as far as I have found, to that moment of sudden snow and the resulting damage perpetrated onto the lunar environment and its inhabitants.

The association between interspecies violence and climate should be considered as the kernel of the movie's dramatic structure. The sudden snow, dramatically speaking, is the actant that catalyses the encounter of species and their civilisations, which inevitably ends with the murder of several Selenites, including their leader. In other words, nothing other than adverse climatological conditions gives the pretext for the violence enacted in the film. Méliès's tale is one of climate politics and conflict. The retreat of the scientists away from the snow leads them to meddle with the lunar biosphere, leading to a defensive reaction from the Selenites, who are reasonably angry. This then leads to the Selenite king being murdered while another lunar individual becomes part of a spectacle back on Earth. Interspecies violence is enacted here by the subjectivity less adjusted to the climatological conditions, the one with lesser knowledge of the land, the one who presupposes its universality.

Italian philosopher Giorgio Agamben famously explains how, at the foundations of Western political thought, there is a connection between the sovereign, who can kill without punishment, and the sacrificed, whose killing is a socially sanctioned act. He writes that:

> The political sphere of sovereignty was thus constituted through a double exclusion … which takes the form of a zone of indistinction between sacrifice and homicide. The sovereign sphere is the sphere in which it is permitted to kill without committing homicide and without celebrating a sacrifice, and sacred life – that is, life that may be killed– is the life that has been captured in this sphere.
>
> *(Agamben 1998, 53)*

Drawing from Agamben's biopolitical thesis, extraterrestrial biopolitics can be defined by technoscience's control on the conditions by which life gets to die outside of this planet. The universe is a theatre, and humans stage death all over it. We rehearse lunar climates to deny our extinction. Extraterrestrial biopolitics could then be more accurately defined as the necropolitics of non-human life, which is sent 'out there' to die so we can live. Méliès thus speculated on an extraterrestrial necropolitics that, back then, was yet to come. Today, however, necropolitics is already in action. We can see how, after Agamben, all non-human life in outer space, all extraterrestrial life, is sacrificial. The human Heliopolis Meliés imagined in his film is germane to the later advancement of an interplanetary polis where the non-human is de facto expendable. From Laika and its wonderfully theatricalised death to the dozen fruit flies that live on board the International Space Station, to the first plant on the Moon: the pursuit of outer space has inevitably been linked with the sacrifice of non-human life beyond our atmosphere.[1] Non-human life in 'outer space' is meant to die in the pursuit of figuring out how humans might survive an eventual post-planetary condition (Tabas 2022). In this sense, an extraterrestrial bio/necropolitical act of lunar performances may already be useful as an analytical framework to understand the aestheticisation of climate change as related to the production of bare life.

In today's new space context, extraterrestrial life is bred in connection with and service to neocolonial discourse enacted in space exploration. Narratives of extraterrestrial coloniality resemble the European colonial project in that the goal is ultimately to reproduce an ontological interior—a climate—where the terrestrial power matrix can be reproduced biotechnologically (Treviño 2023; Sloterdijk 2013). However, while earlier terrestrial colonial projects stem from the skewed vision of human exceptionality, terrestrial and extraterrestrial coloniality differ in that the former makes a human difference the grounds for sacrificial life to appear as such, while the latter has the means to produce, sustain and stratify life (and death) technoscientifically. In other words, whereas the terrestrial colonial project aspired to reproduce the European ecosphere elsewhere on the planet, the extraterrestrial colonial project extends the human *bio*sociosphere onto other celestial bodies by means of creating sacrificial life. And while the extraterrestrial colonial project that today has figures as Elon Musk at its centre descends from the European maritime colonial project of previous centuries, it differs from its ancestor in that what is at stake is the pursuit of a transhumanist project to decide who gets to live on Earth and who gets to die off it. And the pros and cons of both positions are still unraveling as we speak: will the lunar colony be full of expendable life? Or will Earth, now a different planet, be left for miners to produce the means needed by those who can fly away?

Towards Lunar Performance Studies

The actual sustenance of extraterrestrial artificial climates on the Moon and beyond may produce a continuous and likely negative impact on climates back on Earth. What can we learn from a performance history, or perhaps more accurately, a performance account of lunar climates? We can learn about the history of territorialising the solar system. We can learn about the collective impetuses to bring forth replicas of our planet, in a sort of wicked twice-behaved behaviour. A lunar perspective in the conversations of dance, theatre and performance studies ought to consider and problematise the ongoing industrialisation of cislunar and lunar space. Performance studies need an extraterrestrial theory. Other lunar scenarios are urgent, and this is where performance's worlding (and worldly!) power becomes ever so relevant.

Note

1 Olesya Turkina explains that the Soviet Union hid the fact that Laika would die in the Sputnik 2 mission because there was a growing international concern about the fact that dogs were being sacrificed as part of the Soviet space program. 'Sputnik 2 was launched before the scientists jad learned how to recover "the passengers". Although they knew from the outset that Laika would die, official reports knew from the outset that Laika would die, official reports contained no information regarding the dog's

fate' (2014, 86). Laika was officially declared death after eight days in orbit, and several versions circulated regarding the reason for her demise, many of which were variations of the version in which she was administered an euthanasia drug. However, the fact is that Laika suffocated within hours after the launch due to a thermal conductivity miscalculation. For the following week, the official story was that she was alive, and 'newspapers would periodiacally publish reports about her health' (89).

References

Acket-Goemaere, Alizée, Ryan Brukardt, Jesse Klempner, Andrew Sierra and Brooke Stokes. 2024. 'Space: The $1.8 Trillion Opportunity for Global Economic Growth.' *McKinsey & Company*, 8 April. https://www.mckinsey.com/industries/aerospace-and-defense/our-insights/space-the-1-point-8-trillion-dollar-opportunity-for-global-economic-growth, accessed 24 April 2025.

Agamben, Giorgio. 1998. *Homo Sacer: Sovereign Power and Bare Life*. Translated by Daniel Heller-Roazen. Stanford, CA: Stanford University Press.

Clormann, Michael and Nina Klimburg-Witjes. 2022. 'Troubled Orbits and Earthly Concerns: Space Debris as a Boundary Infrastructure.' *Science, Technology, & Human Values* 47 (5), 960–85. https://doi.org/10.1177/016224392110235.

Fox, Alex. 2019. 'China Sprouts Cotton Plants on the Moon.' *Science*, 15 January. https://www.science.org/content/article/china-sprouts-cotton-plants-moon, accessed 2 December 2024.

Lefebvre, Thierry. 2011. 'A Trip to the Moon: A Composite Film.' In Matthew Solomon, ed, *Fantastic Voyages of the Cinematic Imagination: Georges Méliès's Trip to the Moon*. Albany, NY: State University of New York Press, 49–63.

Méliès, Georges, dir. 1902. *A Trip to the Moon*. Manufacture de Films pour Cinématographes.

Scharmen, Fred. 2017. 'Highest and Best Use: Subjectivity and Climates Off and after Earth.' *Journal of Architectural Education* 71 (2), 184–96. https://doi.org/10.1080/10464883.2017.1340775

Scharmen, Fred. 2021. *Space Forces: A Critical History of Life in Outer Space*. London: Verso.

Sloterdijk, Peter. 2013. *In the World Interior of Capital: For a Philosophical Theory of Globalization*. Trans. Wieland Hoban. Cambridge: Polity Press.

Tabas, Brad. 2022. 'Incipit Gaia the Grotesque Garlanded with Satellites: Re-Imagining the Anthropocene as a Post-Planetary Age.' *Cosmos and History: The Journal of Natural and Social Philosophy* 18 (1), 131–49.

Taylor, Diana. 2003. *The Archive and the Repertoire*. Durham, NC: Duke University Press.

Treviño, Natalie B. 2023. 'Coloniality and the Cosmos.' In Juan Francisco Salazar and Alice Gorman, eds, *The Routledge Handbook of Social Studies of Outer Space*. London: Routledge, 226–37. doi.org/10.4324/9781003280507-21

Turkina, Olesya. 2014. *Soviet Space Dogs*. Translated by Ina Cannon and Lisa Wasserman. London: FUEL Design & Publishing.

9

RE-STAGING CABINETS OF CURIOSITIES IN THE AGE OF ANTHROPOCENE

Cen Liu

Cabinets of Curiosities as Ecological Performances

Imagine encountering a jar of sand from a North Carolina beach that contains the layered history of earth cycles and human development, a recording of a now-extinct bird's song mimicked by a human voice, or fabricated fossils and minerals made from plastics, aluminium and bones. These objects serve as portals between present and future, inviting us to witness how human activities have and will have altered the earth. Objects as such have unique import in the age of the Anthropocene, a term coined by palaeoecologist Eugene Stoermer and atmospheric chemist Paul Crutzen to describe a new geological epoch in which humans as geological agents reshape the planet's systems. This critical framework of thinking and action has prompted artists to collect, display and fabricate objects of curiosity to rethink the human relationship with the more-than-human world.[1]

This chapter examines how contemporary cabinets of curiosities function as ecological performances that respond to our ecological crisis. I argue that these projects share fundamental strategies with ecological theatre—performance practices that reject anthropocentrism and instead foreground interconnections between humans and environmental systems, while acknowledging the agency of nonhuman elements (Chaudhuri 1994). Both contemporary cabinets and ecological theatre grapple with presenting nonhuman agency and planetary-scale processes that typically exceed representation. By analysing artist Yesnia Thibault-Picazo's 'The Cabinet Anthropogenic Specimens' and the 'Anthropocene Slam', an event held at the University of Wisconsin-Madison, I demonstrate how contemporary cabinets reimagine traditional display practices as material encounters with planetary transformation.

DOI: 10.4324/9781003584445-12

The medium of theatre offers unique insights into our environmental crisis precisely because it has traditionally been burdened with human-centrism, which makes its ecological reorientation particularly significant. Chaudhuri (2015, 2) argues that theatre now faces a central challenge: addressing the incommensurability between vast planetary forces and limited human systems of thought. This epistemic crisis has given rise to what she calls a 'drama of bad ideas' (10), which deliberately attacks its own anthropocentric foundations, destabilising both ideas (meaning) and forms (stage directions and dialogue). Chaudhuri's observation resonates with Lavery's concept of 'weak performance'—theatre that embraces 'everything that Western modernity distrusts—the weak, the unfinished, the superfluous, the contingent' (2016, 233). Such deliberate vulnerability offers a counterpoint to the mastery mindset that has characterised modernity. Both scholars recognise that conventional theatrical forms have historically perpetuated the very thinking that produced our climate crisis and widespread species extinction.

In this chapter, I argue that contemporary cabinets of curiosities function as ecological performances that contest the very epistemological foundations that produced the Anthropocene through three specific theatrical strategies: first, by disrupting conventional temporalities to make geological time perceptible; second, by inverting traditional power dynamics between human observers and nonhuman materials; and third, by staging embodied encounters that reveal our entanglement with environmental systems. Unlike historical cabinets that reinforced human mastery through taxonomic display, these artistic interventions employ performance strategies from ecological theatre including 'slow dramaturgy', 'deep dramaturgy' and 'hybridity' to transform static display into dynamic engagement. Viewing these cabinets through a theatrical lens reveals how they not only critique anthropocentric worldviews but also rehearse alternative modes of human-nonhuman relations necessary for navigating the current era.

Re-Staging Human-Nonhuman Relationships in the Age of Anthropocene

A curiosity cabinet contains exotica, relics, antiquities, oddities and finely crafted works that evoke wonder through their rarity, craftsmanship or mysterious origins. These collections emerged in early modern Europe as more than assemblages of treasures. Rather, they were epistemological frameworks that shaped human relationships with the natural world. Within these ornate wooden cases and study rooms, collectors tested, dissected and distilled natural specimens, establishing what Findlen calls 'laboratories of nature' (1994, 154). With the rise of Baconian empiricism, strange objects provided the material basis for the taxonomic ordering of nature (Daston and Park 1998, 215–21). As 'social device[s]', these cabinets helped collectors navigate aristocratic circles and intellectual networks (MacGregor 2007, 66). In their classification of specimens reduced to collectible curiosities, these historical cabinets embodied what Joseph calls the Enlightenment's 'grand narrative[s] about human origins'—narratives that positioned European collectors as masters of a world increasingly subjected to their gaze (2018, 29).

The cabinet form's historical complicity in establishing the conceptual foundations of the Anthropocene makes it a telling site for critical intervention. Early modern cabinets instituted three epistemological frameworks that enabled the extractive relationships characterising our current ecological crisis. First, they normalised the conceptual separation between humans and nonhumans, positioning nature as passive material awaiting human interpretation. Second, they established a taxonomic approach to the natural world that fragmented ecological relationships into discrete, collectable specimens. Third, they privileged visual knowledge and ownership over other forms of sensory engagement. Contemporary artists work within and against these problematic frameworks, exposing these conceptual foundations while imagining alternative relationships.

Thibault-Picazo's ongoing project 'The Cabinet of Anthropogenic Specimens' (initiated in 2012) offers a rich case study. Unlike traditional cabinets of curiosity that displayed existing natural specimens as possessed objects of knowledge, Thibault-Picazo's cabinet creates speculative future specimens that materialise the planetary impact of human activity. Developed in collaboration with geologist Dr. Jan Zalasiewicz, the cabinet presents what Thibault-Picazo calls 'an evolving material library representing future geology' (2025). This framing inverts the temporal orientation of the cabinet of curiosity. Where historical cabinets claimed to comprehensively represent existing knowledge, Thibault-Picazo's cabinet performs the incompleteness of our current understanding, projecting forward to imagine what knowledge might be available to future observers of our present actions.

Thibault-Picazo's work confronts the same problem that Chaudhuri identifies in ecological theatre—that of scale. More specifically, cabinets of curiosities reveal the limited scale of human experience of time in the age of Anthropocene, as geological processes cannot be measured by our biological process. 'The Cabinet Anthropogenic Specimens' brings tangible shapes to this limitation through the creation of unnatural specimens: hybrids, mutations and synthetic-organic amalgams. These specimens undo the taxonomic scale of the traditional cabinet, amplifying the temporal disjunction by showing how our everyday objects might appear as archaeological remnants. In this sense, the objects embody what Timothy Morton calls 'hyperobjects'—phenomena so massively distributed across time and space that they exceed immediate comprehension (2013, 1). For example, the Cumbrian Bone marble, made from the bones of sheep and cattle slaughtered during the 2001 foot and mouth epidemic, materialises the environmental consequences of industrial agriculture. The Pacific Plastic Crust mortar visualises how ocean plastic pollution will transform into geological strata. The aluminium vessel reflects the industrial waste dumped into waterways. In this way, Thibault-Picazo's work makes palpable form to these otherwise abstract environmental processes.

Further, the cabinet becomes a dramaturgical device that allows human viewers to witness geological processes that normally exceed human lifespans, making slow planetary changes perceptible through their artificial acceleration. Thibault-Picazo employs what she terms 'geomimicry', a methodology that mimics and accelerates geological processes by using heat presses to apply extreme pressure to her materials. Eckersall and Paterson's concept of 'slow dramaturgy'—a practice that shows 'awareness of time as an aesthetic-corporeal medium of expression'—helps illuminate the ecological implications of this method (2011, 183). In Thibault-Picazo's cabinet work, slow dramaturgy manifests through accelerated processes—her heat presses compress geological time. This apparent contradiction is precisely the point: by radically altering temporal experience, she creates the conditions for the reoriented corporeal experience of everyday-ness that slow dramaturgy seeks to achieve.

Thibault-Picazo does not position her cabinet as offering solutions to the environmental crisis. Instead, she states that her work aims 'to provoke a reflexion around our conception of nature and ultimately to invite to reimagine how we inhabit the world'. This approach aligns with what Stevens defines as 'deep dramaturgy', that is, a practice that 'demands we surrender human exclusivity over artistic production' (2019, 96). The work echoes this deep dramaturgy as the artist does not order or categorise the objects. Instead, Thibault-Picazo emphasises ongoing processes of material transformation, describing her cabinet as 'evolving', with each collection 'meticulously developed' through site-specific research into 'historical and natural heritage' with 'particular emphasis on current industrial activities' (2025). While offering this curation, she also fosters incompletion in human viewers' encounters with the cabinet.

The 'Anthropocene Slam: A Cabinet of Curiosities' project further exemplifies how the cabinet form can be activated through performance into a site of embodied encounters. Organised by the Center for Culture, History, and Environment at the University of Wisconsin-Madison in collaboration with the Rachel Carson Center for Environment and Society in Munich and the Environmental Humanities Laboratory at KTH Royal Institute of Technology in Stockholm, this event took place in 2014. It brought together artists, anthropologists, historians, geographers, literary scholars and biologists to present objects that might form a cabinet for the Anthropocene. Unlike a conventional academic conference, the slam adopted the format of a poetry slam, with participants performing their objects in a series of freestyle, timed presentations. Objects ranged from concrete to origami passenger pigeons, from recordings of extinct bird calls to everyday plastic items. The slam's organisers consciously chose the cabinet structure as their framework, drawing on its historical associations with knowledge-ordering while altering its function. As Mitman et al. explain, the cabinet structure allows objects to 'disrupt linear narratives, including those about the Anthropocene' by bridging 'spaces and join[ing] times' and 'summon[ing] all at once the past, present, and future' (2018, xi). The slam reimagines the cabinet as a performative space where objects and their stories can unfold through time and across bodies. Rather than treating objects as dead specimens for human examination, the slam adopts a more collaborative, materially responsive approach to knowledge creation.

Parker-Starbuck's concept of hybridity offers a framework for understanding the performances in this event. She defines hybridity as more than simple mixture of elements, approaching it as 'a way to provoke a different "way of seeing" interspecies relationships that "attune the 'human' to the other lives in its periphery"' (2020, 27). Her conception moves beyond fixed biological definitions to emphasise hybridisation as an ongoing process of becoming that creates 'networks of collective interspecies imaginings' (28). In the origami passenger pigeon activity, the performance enacts a form of engagement where participants physically connect with the memory and form of an extinct species by literally creating the paper bodies of the birds. As the paper birds take flight across the performance space, the humans experience an embodied sense of how animal lives impact our own, fostering a hybrid type of consciousness through the imaginative process of folding, creating and animating the birds. In the concrete pouring performance, the performers mix raw concrete onstage, allowing audiences to witness its transformation from dry material to fluid mass to solid form. This process reveals concrete's agency—its ability to both conform to human intentions while simultaneously asserting its own material properties through unpredictable flows, textures and curing patterns. The performance thus enacts a hybrid relationship where humans and material collaborate in creation. Through this, participants can recognise concrete not simply as an inert building substance but as a geological actor with its own forms of agency.

The slam events were later documented in a volume titled *Future Remains: A Cabinet of Curiosities for the Anthropocene*, which expanded the cabinet concept beyond the initial performance. This publication itself represents a yet inverted temporal disruption by transforming ephemeral performances into a more lasting archive while still emphasising the provisional nature of knowledge in the age of Anthropocene. The book's organisation reinforces the cabinet's disruption of power dynamics. The chapters are arranged in what Mitman et al. describe as 'a series of reflective essays organised around 15 exemplary objects that offer a fragmentary history of the Anthropocene' (2018, xiii). This fragmentary approach resists comprehensive mastery, instead inviting readers to 'browse, dip in, and explore' as participants in making sense of the environment that they are entangled in (xiii).

Conclusion

Contemporary artistic adaptations of cabinets of curiosities confront the structures of knowledge that separate human observers from nonhuman materials—the subject-object distinctions that have enabled environmental extraction. A theatre and performance lens to cabinets is generative in that it reveals how their contemporary shift to relationality and processes destabilise the epistemological frameworks of human supremacy. In an era when climate change collapses distinctions between human and geological timescales, the cabinet form offers a framework that accommodates multiple temporalities simultaneously: it makes visible the processes through which the present becomes the past and/or the future, allowing audiences to experience the entanglement of different time scales through material encounters. By engaging multiple senses including sight, sound and touch, the contemporary cabinet also inverts the power hierarchy between human viewers and objects, anchoring knowledge making in embodied encounters. Thus, these cabinet performances offer critical interventions that expose the epistemological foundations of the current ecological crisis while rehearsing alternative modes of human-nonhuman relations.

Note

1 This includes the research by artist Kelly Jazvac and earth scientist Patricia Corcoran on plastiglomerates, pebbles, shells and rocks that contain artificial debris. 3D designer and animator Louise Silfversparre's project *Technofossils* features a series of 3D-rendered animations telling the stories of six minerals that are entangled with human activity. Outside the gallery, Alice Twemlow's *Design and the Deep Future* programme at the Royal Academy of Art, the Hague (KABK) encourages staff and students to investigate how designers can create objects that cause less damage in the process of fossilisation. A workshop and graduate summer school in Canada/United States titled *A Museum for Future Fossils* investigates how curatorial practices can prompt us to think about human impact on the environment.

References

Chaudhuri, Una. 1994. '"There Must Be a Lot of Fish in That Lake": Toward an Ecological Theater.' *Theater* 25 (1), 23–31. https://doi.org/10.1215/01610775-25-1-23.

Chaudhuri, Una. 2015. 'Anthropo-Scenes: Theater and Climate Change.' *Journal of Contemporary Drama in English* 3 (1), 12–27. https://doi/10.1515/jcde-2015-0002.

Daston, Lorraine and Katherine Park. 1998. *Wonders and the Order of Nature: 1150–1750*. New York: Zone Books.

Eckersall, Peter and Eddie Paterson. 2011. 'Slow Dramaturgy: Renegotiating Politics and Staging the Everyday.' *Australasian Drama Studies* 58, 178–92.

Findlen, Paula. 1994. *Possessing Nature: Museums, Collecting, and Scientific Culture in Early Modern Italy*. Berkeley: University of California Press.

Joseph, Betty. 2018. 'Worlding and Unworlding in the Long Eighteenth Century.' *Eighteenth-Century Studies* 52 (1), 27–31.

Lavery, Carl. 2016. 'Introduction: Performance and Ecology – What Can Theatre Do?' *Green Letters* 20 (3), 229–36. https://doi/10.1080/14688417.2016.1206695.

MacGregor, Arthur. 2007. *Curiosity and Enlightenment: Collectors and Collections from the Sixteenth to the Nineteenth Century*. New Haven, CT: Yale University Press.

Mitman, Gregg, Marco Armiero and Robert S. Emmett. 2018. 'Preface.' In Gregg Mitman, Marco Armiero and Robert S. Emmett, eds, *Future Remains: A Cabinet of Curiosities for the Anthropocene*. Chicago, IL: The University of Chicago Press, ix–xiv.

Morton, Timothy. 2013. *Hyperobjects: Philosophy and Ecology after the End of the World*. Minneapolis: University of Minnesota Press.

Parker-Starbuck, Jennifer. 2020. 'Becoming-Leech.' *Performance Research* 25 (4), 26–35. https://doi/10.1080/13528165.2020.1842027.

Stevens, Lara. 2019. 'Anthroposcenic Performance and the Need for "Deep Dramaturgy".' *Performance Research* 24 (8), 89–97. https://doi/10.1080/13528165.2019.1718436.

Thibault-Picazo, Yesnia. 2025. 'The Cabinet Anthropogenic Specimen.' https://www.yeseniatp.com/THE-CABINET-ANTHROPOGENIC-SPECIMENS, accessed 1 May 2025.

Environmental Violence and Inequity

10

'SLOW VIOLENCE' IN CHERRÍE MORAGA'S *HEROES AND SAINTS*

Elin Diamond

The job of an artist is to anticipate what academics will be thinking about.

—*Cherríe Moraga*

Cherríe Moraga is a writer of my generation, one whose political transformations were partially fuelled by second-wave feminism in the United States. She was born in Los Angeles in 1952 to a mother who provided her with strong roots in Mexican culture. By the late 1960s and 1970s, California's Mexican-American communities began morphing into an activist Chicano/a culture which, with Indigeneity's pull, was later refigured as *Xicano/a,* a potent combination of ancient Aztec myth (Aztlán) and the contemporary urban street. These dynamic refigurings gathered in Moraga's other identities, including that of *mestiza* (her father was white), while Moraga's lesbian identity was nurtured by the burgeoning feminist movement of the 1970s. However, while grateful for the consciousness-raising and feminist solidarity that helped her survive the endemic machismo of Chicano culture, she found Bay Area feminism too white and too thoughtless about Chicano political thought with its la Raza dreams and culture-building. In the 1980s and 1990s, Moraga built a playwriting career and became an activist Xicana artist, essayist, and professor, based for 20 years at Stanford University and now at the University of California, Santa Barbara (Moraga 2011, 2017). I've introduced her twice, once long ago to a large, friendly academic audience and a second time more recently to a group of sceptical Latino/students at Rutgers University for whom Chicana radicalism of the 1970s and 1980s belonged to the last ice age.*

*Ice age. I am thinking about climate change and dramatic form. Ecological problems are massively overdetermined, involving multiple causes accumulated

DOI: 10.4324/9781003584445-14

and distributed across time and place. They do not lend themselves to examination through the reassuring causality of a plot. Yet, disaster tourists take trips to Arctic regions imagining they'll witness a chunk of glacier fall thrillingly into the sea while their cameras are poised to capture it, after which a swift denouement—safe return to a heated hotel room. Ecological disaster may require something like a baggy multi-temporal episodic form in order to register the many facets of climate change. Perhaps we need long-form performances that tell mythic stories which, while absorbing the energy of theatre's Liveness, can also tap a vein of cultural memory *and* futurity, of family, of familia.*

*Familia is the connecting thread to the myth of Aztlán, an idea forcefully reinvented by Chicanx activists in the 1960s and 1970s out of rage at the continued effects of colonial oppression. 'According to myth, Aztlán was the ancestral homeland that the Aztecs left in 1168 when they journeyed southward to found ... Tenochtitlán [Mexico City] in 1325' (Arrizón 2000). Ancient codices testify that Indigenous founders of Tenochtitlán came from El Norte—the region that encompasses what is now the southwestern United States. After Columbus, these regions were taken/stolen and annexed, and after the Mexican-American War (1846–48), Mexico signed treaties that ceded to the United States the territories that became Texas, New Mexico, Arizona, California, and parts of Utah and Colorado. Mexicans living in these places were entitled to continue living there, working the land. But these treaties were violated or ignored soon after the ink was dry. Mexican and Indigenous people who had farmed for generations became second-class citizens, not 'real Americans'. Thus, for Moraga and her fellow theatremakers from the late 1960s to the 1990s, Aztlán asserted a *precolonial* identity, an Indigenous Aztec identity that names Mexicans as origin people, as familia, whose bonds precede, and exceed, national citizenship. Writes poet Gloria Anzaldua in 'The Homeland, Aztlán' (1987, 3):

> This land was Mexican once,
> was Indian always
> and is.
> And will be again*

*Again. Always. Will be. Such is the temporal grammar of political myth. Suffering injustice in the present, a desperate people asserts the continuity of an unremembered cultural past: Aztlán *was* and *is always* the present. By reclaiming its roots and rituals a community converts desperation into strength and prepares for the future. This is futurity without a plot, a promise without guarantees, a dreamscape and also a landscape.*

Landscape introduces the human eye which claims and holds the natural world in its human-sized frame. Aztlán, by contrast, is about the land as an unframed continuous Indigenous area that existed before and continues to exist, if only in reverie, after the Spanish, Mexican and US land grabs. Aztlán is about scale. With Aztlán in her sights and soul, the individual joins a great forever of Indigenous generations, who themselves belong to the seemingly forever of animals, plants, sky and water. For Xicanx artists, scholars and activists, Aztlán connotes both a homeplace and a healthy ecology underpinned by cultural knowledge and respect for the land (May 2021). It asserts the ecological congruency of the land and its rivers, deserts, species migration and all the ways it has provided human sustenance.

*Sustenance, what sustains us? The real and fantastical myth of Aztlán yawns over Moraga's imagination, but because she's a feminist, living in her moment, history intersects myth, often brutally. Produced by a Xicana company, Brava! For Women in the Arts, in San Francisco in 1992, *Heroes and Saints* is set in McLaughlin, a fictional town off Highway 99, but it is based on the verified pesticide poisoning of agricultural workers in an actual town, McFarlane, CA, over a ten-year period, 1978–88, in California's rich San Joaquin Valley. Pregnant field workers, picking fruit and vegetables, passed toxins to their developing babies, many of whom were born with horrendous birth defects—bodies without limbs or vital organs. At that time, childhood cancer rates in McFarlane were 400 per cent above normal* (Chávez 1989; May 2021, 217).

*Normal? What is normal when, generation after generation, you're treated like a work animal, your children disposable? Aggregating elements from Cesar Chávez's United Farm Worker struggles (the first grape boycott [1965–70] and the second—anti-pesticide—boycott in the 1980s), Moraga draws inspiration from the UFW's stunning documentary of 1986, *The Wrath of Grapes*, which documents the terrible deformities in infants and young children born to pesticide-infected mothers (UFW 1986). Cerezita, the protagonist of *Heroes and Saints*, is a pesticide baby. Born without a body, she is now a teenager and an astute analyst of political hypocrisy. She holds forth while perched on a table with wheels, which she operates by pushing buttons with her tongue and chin ('*The hum of [her motorized rolling table] always anticipates her entrance*', Moraga 1994, 90). Her farm worker mother, Dolores, ashamedly hides Cerezita behind curtains; her hairdresser sister Yolanda breastfeeds her newborn while doing her sister's hair; her gay brother Mario loves his family but can't live with them—he's a Valley rent boy, earning money as a male prostitute to gringos driving down Hwy 99, and shows symptoms of AIDS. The timid local priest, Father Juan, is in thrall to the intelligence of Cerezita and struggles with his own desires; TV reporter, Ana Perez, is a satirical portrait of a media witch, and Amparo, Dolores's *comadre*, co-worker and friend, is Moraga's respectful portrait of the indefatigable Dolores Huerta, who co-founded the United Farms Workers union with Cesar Chávez.*

*Chávez, admired by Martin Luther King, Jr. and the political radicals of my growing-up years in Berkeley, dominated the political landscape. But the emotional centre of *Heroes and Saints* is the Valle family who live in flimsy company-built housing, surrounded by

> hundreds of miles of soil… This soil should not be confused with land. What was once land has become dirt, overworked dirt, overirrigated dirt, injected with deadly doses of chemicals and violated by every manner of ground- and back-breaking machinery. The people that worked the dirt do not call what was once the land their enemy. They remember what land used to be and await its second coming.*
>
> *(1994, 91)*

*Second coming. Aztlán is saturated with spiritual longing. The vast majority of Mexican-American farmworkers are Catholics, and though there is irony directed at Father Juan, the people embrace him because church practices and rituals foster precious community. However, Moraga's fictional town of McLaughlin has created its own rituals for honouring Christ's birth and crucifixion. When infants die of pesticide poisoning, McLaughlin's activists ask parents not to bury them where they will be forgotten. Instead they remove the bodies from their little coffins and hang them on crude crosses that are planted in the vineyards alongside the grapes that are poisoning the workers who pick them, and who give birth to more deformed children. For Moraga (as for Marx), capitalism in the life of a farm labourer is a cycle of death. In the opening stage directions, Moraga unites the bodiless Cerezita with an anonymous dead child, both becoming Christological symbols:

> At rise in the distance, a group of children wearing calavera [skull] masks enters the grape vineyard. They carry a small, child-size cross which they erect quickly and exit, leaving its stark silhouetted image against the dawn's light. The barely distinguishable figure of a small child hangs from it. The child's hair and thin clothing flap in the wind. Moments pass. The wind subsides. The sound of squeaking wheels and a low, mechanical hum interrupt the silence. CEREZITA enters in shadow. She is transfixed by the image of the crucifixion. The sun suddenly explodes out of the horizon, bathing both the child and CEREZITA. CEREZITA is awesome and striking in the light. The crucified child glows, Christlike. The sound of a low-flying helicopter invades the silence. Its shadow passes over the field. Black out.
>
> *(1994, 92)*

The low-flying helicopter delivers pesticides throughout the play and at the end, something worse. When Cerezita insists on throwing off the curtains her mother shrouds her with, she and Father Juan lead a procession to the fields for the crucifixion of her sister's dead baby in defiance of the threats of the growers. The sound of a low-flying helicopter terrifies the crowd, who drop to the ground. This time the growers' hired men are spraying not pesticides but bullets.*

*Bullets, extreme violence, murder, terror, this is the currency of social injustice poisoning our moment in US history, global history and the Anthropocene, and to which the poor and the habitually exploited are especially vulnerable. Una Chaudhuri has urged that environmental theatre be an 'assault on the traditional divisions, taxonomies, and conventions of theatre' (1995, 23). Audience separation in a darkened space is indeed a classic convention. Moraga, however, makes use of the audience divide while naming her audience EL PUEBLO, that is, 'THE PEOPLE/PROTESTERS/AUDIENCE … ideally it should be made up of an ensemble of people from the local Latino community'. Amiri Baraka, also writing for his local community at Spirit House in Newark, NJ, forcefully exploded the conventions of theatre but always held onto its magic, both in writing and production. Baraka's stage directions are prose-poems conveying '*smells and cries … a total atmosfeeling*' (Baraka 1978, 132). Moraga's images are similar, while carrying her own political bite. *Heroes and Saints* draws from the satiric *actos* in the style of Luis Valdez's Teatro de Campesino—agitprop scenes mocking, for example, racist stereotypes (see Moraga's pandering reporter Ana Perez)—and the longer form *mitos* (myths), stories of Aztlán spirituality and ecological resilience (Huerta 2000). Yet, she also belongs to her generation of engaged theatremakers of the 1970s and 80s, including Maria Irene Fornes, Baraka, Caryl Churchill, David Hare and, closer to home, the collective of El Teatro de la Esperanza (Yarbro-Bejarano 1990, 140–42), all of whom, in highly varied ways, experiment with flexible forms in which an overarching political narrative is draped over lively short scenes in multiple linguistic and historical registers.*

*Registers. I like this word. In the singular, it's an official record book. But the plural points to musical tones and clefs, which sociolinguists tie to situations. Multiple registers are vital if playwrights are going tell—and merge—complex ecological stories, encompassing not just generations but also centuries. As noted above, such plays sideline plot, with its baggage of reassuring causality. Instead the emphasis is on gesture, image, mood and playful address to the audience. Moraga's *Heroes and Saints* doesn't seek to tell a universal story but rather a specific story of the suffering of a family of Xicana women working in a poor company town in the richest agricultural area in the world. Formally and linguistically, Moraga's play explodes the confines of McLaughlin to become a greater generational story of racial oppression in a poisoned world produced by extractive capitalism. Crucial to *Heroes and Saints* are the many registers of code-switching, signalling migration from vernacular English to vernacular Spanish and back again. Some of my New Jersey students hated this about

Moraga's play—'it's not fair', they said, 'this is an English department class and our text isn't in English'. But most understood the underlying politics of a queer Xicana writing for a theatre where the power brokers and most of the audiences, even in BIOPC neighbourhoods, are Anglo and comfortable interpreting a world they mostly control. If Spanish temporarily confounds the reader and listener, Moraga helps by writing dialogue wherein one character's Spanish seamlessly becomes another character's English, with none of the overt markers of translation. Interestingly Cerezita, speaking from her rolling mechanical table, never code-switches. She recites scripture in flawless Spanish to the priest, and a powerfully imagistic English ignites her last monologue. Whether at the theatre or in class, no linguistic group is entirely at ease. Moraga's text forces us to listen carefully, to go slow.*

**Slow Violence and the Environmentalism of the Poor* is Rob Nixon's attempt to capture the time-scale of climate change as well as today's sense of constant crisis. Violence, he points out, is not only exploded buildings and bodies, daily shootings and blood-strewn streets but also 'a violence that occurs gradually and out of sight, a violence of delayed destruction that is dispersed across time and space, an attritional violence … incremental and accretive' (2011, 2). As theatre scholars we ask how to represent the slow violence of climate change, the slow violence of toxins soaking into the soil that human beings cultivate for wealthy agribusinesses and for the sustenance of their dinner table. In other words, how does one tell the story of the land when ecological disaster occurs over time, sometimes extended historical periods? how can theatremakers make visible the slow violence of climate change, its 'long dyings'* (2).

*Dyings that are long, that are invisible because they are long, stretch the imagination of the dramatist. In *Heroes and Saints*, Moraga turns Yolanda's pregnancy into a Brechtian *gestus*, a pointed contradiction, in which the passage of time for the growth and development of a foetus becomes instead the means to maim and kill it. When we first meet Yolanda she is breastfeeding little Lina. Watching her, the bodiless Cerezita comments: 'I remember the first time I tasted fear; I smelled it in [my mother's] sweat. It ran like a tiny river down her breast and mixed with her milk. I tasted it on my tongue. It was very bitter. Very bitter' (Moraga 1994, 95). Looking at Cerezita's disembodied talking head, we infer immediately that Dolores' milk, mixed with her sweat, was toxic. And in the present time of the play, invisible toxins carried in Yolanda's milk are being swallowed by *her* baby, now growing and multiplying in *her* cells. This is toxicity in slow-release; the disease is invisible, but also continuous, unstoppable. In California's ruthless agribusiness, where a farmworkers' labour—Dolores's, Yolanda's—is purchased, her body abused and then discarded, breast milk, that most intimate fluid usually given freely by a woman to nurture an infant, cannot remain free or untainted. Slow violence is how capitalism works and spreads, and nothing escapes its reach. At the end of the play, the baby is dead and about to be 'resurrected' on her infant cross, and Cerezita will make the agents of agribusiness, surveilling her in their helicopter, shoot and martyr her in public, in front of the television cameras. (For once the TV reporter Ana Perez is looking in the right direction.) And in martyrdom, there is always blood.*

*Blood dominates the images in Cerezita's final speech, beginning with the shocking line:

> Put your hand inside my wound. Inside the valley of my wound, there is a people. A miracle people. In this pueblito where the valley people live, the river runs red with blood; but they are not afraid because they are used to the color red. It is the same color as the river that runs through their veins, the same color as the sun setting into the sierras, the same color of the pool of liquid they were born into....
>
> *(148)*

References to the wound of Christ (at this point Cerezita is dressed as the Virgin of Guadalupe) give way to images of maternal blood, and to a political invocation of Libertad. 'You will be free. Free to name this land *Madre. Madre Tierra. Madre Sagrada.* The radiant red mother...rising' (148). Here, as throughout the play, Catholicism's discourse is infused with the language of nature and the dreamworld of Aztlán.*

*Aztlán again. What is Moraga's purpose in invoking, and romanticising, these distant peoples before colonialism, before western hegemony, before the terrible and ongoing harms of global capitalism? One answer is that Aztlán offers Moraga's audiences an image of ecological healing, of *slow healing*. How does it begin? Recall that in the time of the play, Cerezita and Father Juan have just been gunned down by the growers' thugs. They lie amid the grape vines, while everyone onstage drops to the ground in fear. Then events move fast....

> Mario *rises, raises his fist into the air:* Burn the fields! Then *El Pueblo (rising with him): Enciendan los files! (They all, including [the reporter] Ana Perez rush out to the vineyards, shouting as they exit.) Asesinos! Asesinos! Asesinos! Moments later, there is the crackling of fire as a sharp red-orange glow spreads over the vineyard and the Valle home. The lights slowly fade to black.* End.*
>
> *(149)*

End, without end. 'The crackling of fire' offers El Pueblo a catharsis, or the image of one. Is it satisfying? From the documented pesticide poisonings of the 1980s to our data-saturated contemporary moment, the news is not good. A study published just weeks ago, 16 September 2024, finds that 'ten percent of children and 22 percent of adults in California's San Joaquin Valley may be breathing in detectable levels of pesticides, including chlorpyrifos, an already banned chemical linked to neurological damage in children' (Howard 2024). When corporate profits are in question, extractive capitalism will not be stopped. And so Moraga dips into the storehouse of Aztlán mythography to invent a flamboyant and bizarre ritual, the tying of babies' corpses to homemade crosses on which '[t]he child's hair and thin clothing flap in the wind*'. This is horror as poetry, as theatre. This is how we recognise the people's suffering and how, Moraga suggests, slow healing might begin.

References

Anzaldúa, Gloria. 1987. 'The Homeland, Aztlán.' In *Borderlands/La Frontera: The New Mestiza*. San Francisco, CA: Aunt Lute Books, 3.

Arrizón, Alicia. 2000. 'Mythical Performativity: Relocating Aztlán in Chicana Feminist Cultural Productions.' *Theatre Journal* 52 (1), 23–49.

Baraka, Amiri. 1978. 'Slave Ship: A Historical Pageant.' In *The Motion of History and Other Plays*. New York: Wm Morrow and Co, 130–50.

Chaudhuri, Una. 1995. *Staging Place: The Geography of Modern Drama*. Ann Arbor: University of Michigan Press.

Chávez, Cesar. 1989. 'Statement to Pacific Lutheran University.' Public Address at Pacific Lutheran University, Tacoma, Washington, March. https://ufw.org/research/history/cesar-chavezs-first-major-address-36-day-1988-fast-pesticide-poisoning-farm-workers/, accessed 15 June 2022.

Howard, Lisa. 2024. 'Residents in San Joaquin Valley Breathe Chemical Pesticides According to New Study.' *UC Davis Health News*, 11 September. https://health.ucdavis.edu/news/headlines/residents-in-san-joaquin-valley-breathe-chemical-pesticides-according-to-new-study/2024/09, accessed 1 October 2024.

Huerta, Jorge. 2000. *Chicano Drama: Performance, Society & Myth*. New York: Cambridge University Press.

May, Theresa J. 2021. *Earth Matters on Stage: Ecology and Environment in American Theater*. London: Routledge.

Moraga, Cherríe. 1994. *Heroes and Saints*. In *Heroes and Saints & Other Plays*. Albuquerque, NM: West End Press.

Moraga, Cherríe. 2011. *A Xicana Codex of Changing Consciousness: Writings, 2000–2010*. Durham, NC: Duke University Press.

Moraga, Cherríe. 2017. 'Queer Aztlán: The Reformation of Chicano Tribe. In Rudolgo Anaya, Francisco A. Lomelí and Enrique R. Lamadrid, eds. *Atzlán: Essays on the Chicano Homeland*. Revised and Expanded Edition. Albuquerque: University of New Mexico Press, 223–41.

Nixon, Rob. 2011. *Slow Violence and the Environmentalism of the Poor*. Cambridge, MA: Harvard University Press.

UFW. 1986. *The Wrath of Grapes: Pesticide Poisonings of Grape Workers*. A Production of the Volunteer Staff of the United Farm Workers of America. www.youtube.com/watch?v=Wq48o4ftL4A, accessed 2 December 2024.

Yarbro-Bejarano, Yvonne. 1990. 'The Female Subject in Chicano Theatre: Sexuality, 'Race,' and Class.' In Sue-Ellen Case, ed., *Performing Feminisms: Feminist Critical Theory and Theatre*. Baltimore, MD: The Johns Hopkins University Press, 131–49.

11

URBAN ECOLOGIES IN PERFORMANCE

From Colonialism to Community in Peter Morin's *Cultural Graffiti* and Fast Familiar's *The Strategy Room*

Lisa Woynarski

We are living in the 'urban century' (Gaston 2010) as global patterns of urban migration mean that more than half the world's population now live in cities. They are not mere geographical locations, rather they shape the lives of those who live in and around them – socially, politically and ecologically. They reflect the inequalities of society at large and embody historic nature/culture divides, histories of displacement and dispossession, anthropocentrism and material inequalities (Woynarski 2025).

This chapter brings together two very different performance pieces to consider future ways of living together in cities, in the context of climate crisis. Peter Morin's (Tahltan Nation) *Cultural Graffiti in London* (2013) taps into long and often hidden histories, foregrounding how legacies of colonialism are living in the structures of contemporary cities. By singing songs of Tahltan Nation to culturally significant sites and monuments in London, Morin makes visible the ongoing Indigenous presence of colonialism and its extractivist ghosts. *The Strategy Room* by Fast Familiar (2023a) is a participatory work in which audiences are immersed in a fictional future and need to make decisions about their local city. This performance playfully enacts future ecological narratives of the city, imagining hopeful possibilities, and asking the audience to make decisions about net-zero climate policies. I argue (from the perspective of an urban-dwelling settler Canadian living in the United Kingdom) that considering these two different approaches to performance helps to understand the complexities of urban ecologies, from the ongoing legacy of colonialism, to the potential for community-driven practical climate policy.

DOI: 10.4324/9781003584445-15

Cultural Graffiti in London

2013. Buckingham Palace. Tahltan Nation artist Peter Morin has travelled here from what is now known as Canada to undertake a series of 14 performance inventions at historical sites and monuments across the city. At Buckingham Palace, he is wearing a black and red blanket, with white buttons stitched to form a pattern of a crow wearing armour, in a traditional Tahltan style. He is also wearing handmade high-top moccasins (or hunter moccasins) made by his Aunty Lilly. For this intervention, he wears a black domino mask in the style of the Lone Ranger that includes hand-sewn mother-of-pearl buttons to match his button blanket. He sets himself up directly in front of the gates at Buckingham Palace, he sings directly into the ground underneath their palace, he sings and drums around their monument directly in front of their palace; and when done, he walks over to their gate and marks their gate with these words: 'today you lose everything, today you lose everything, because we remain, we are vibrant, we are Idle No More'. After this action, outside Canada Gate he sings Tahltan songs directly to stones underfoot, to the monuments, to his ancestors—past, present and emerging. For this entire performance, he plays a drum with the words 'this drum supports Indigenous voice' as he sings. He is accompanied by a documentation team of witnesses and surrounded by sometimes curious tourists. After his marking of the Buckingham Palace gate, he turns to walk away from the palace. A police officer stops him to interrogate his previous actions, but the officer is eventually satisfied with the explanation of Morin being a university researcher who is researching the effect of British colonialism on Indigenous bodies. He does not detain him. Morin sings at other monuments across Greater London, including sites with explicit Indigenous connections such as Pocahontas' gravesite at St George's Church, Gravesend and Mungo Martin's Totem Pole (which was carved in Canada, after the Potlatch ban was lifted in 1953 as a gift to Queen Elizabeth) that now resides in Great Windsor Park.

Colonialism shaped the ecological relationships of the city and continues to exert ideological control. Historian Coll-Peter Thrush asserts that Indigenous people's presence in London is not 'hidden histories' but forced forgetting (2016). That in fact, when Indigenous visitors/captives or diplomats from North America, New Zealand/Aotearoa and Australia were in London, they were hyper-visible because of their difference. Treated as curiosities, performers, 'noble savages' and/or slaves, Thrush contends they were some of the most visible people in London from the sixteenth to the twentieth century. Morin, dressed in Tahltan traditional dress (along with dress pants and a collared shirt purchased from Marshalls) was hyper-visible during his performance acts, inciting curiosity and suspicion (including a police confrontation). The sonic/aural graffiti performed by Morin was delivered to more-than-human actants, but it was necessary to negotiate London as a highly surveilled and demarcated space. The commissioners of the work, theatre scholars Gilbert and Phillipson, write that

> While the main "audience" for these performances was the site of the song – the ground, the stones, the artwork, the traces of ancestors remaining – Morin felt it was important to include witnesses to his work. The immediate role of these individuals was to document the performances and support the artist by negotiating the everyday practical challenges of working in London's public spaces.
>
> *(Gilbert and Phillipson 2015, 15)*

While performing, Morin overheard observers at these busy tourist sites say 'I think he thinks he's an Indian' and 'shhh … this is an Indigenous performance' (Morin in Robinson 2019, 229). For Indigenous sound scholar Dylan Robinson (Stó:lō/Skwah) who was documenting the work, Morin was enacting Indigenous healing, sovereignty and kinship through his performance. Though this wasn't necessarily readable to the public of tourists visiting the sites and moments because of the lack of visible colonial and Indigenous histories.

Morin is an artist, curator and professor at the Ontario College of Art and Design, Toronto. Morin's piece was commissioned for the international exhibition of Indigenous arts and performance, *Ecocentrix: Indigenous Arts, Sustainable Acts*, in London as part of the Indigeneity in the Contemporary World Research project led by Helen Gilbert. Morin describes his cultural graffiti, singing the songs of Tahltan Nation where the monuments meet the land, 'as a way of embodying Tahltan knowledge, embodying a way of being in relation to people, way of being in relation to the land here…trying to enact a direct challenge to the colonial power' (Morin in Huarcaya and Phillipson 2013). Morin, in a mac jacket and a beaded hide vest made by Tahltan artist Penny Louie, sings Tahltan Songs composed by Tahltan Elder Beal Carlick at historical monuments and places around the city: British landmarks, including the Houses of Parliament, Buckingham Palace and Big Ben; Indigenous monuments such as Pocahontas' gravesite and Kwakwaka'wakw carver Mungo Martin's Totem Pole in Great Windsor Park, Southwark Cathedral; and contemporary pilgrimage sites like the Princess Diana Memorial Fountain in Hyde Park and a Banksy in Brixton. Here, Indigeneity is a transportable concept, which can be enacted as a way of thinking and set of practices, on to different geographies and urban centres. Morin's singing takes on a different meaning depending on the site: at the landmark sites, it calls forth buried Indigenous presence erased in the monuments of empire. At the Indigenous sites, he sings in solidarity and kinship. Indigeneity is enacted through singing and ceremony, speaking to ancestors past, present and emerging, cutting through the colonial legacy of Indigenous erasure. Anishinaabe geographer Heather Dorries writes that the urban can be both liberatory and discriminatory for Indigenous peoples and that performance can play a role in its liberatory potential by articulating urban Indigenous identity (2023, 144). Through Indigenous aesthetics, Morin's work asserts Indigenous presence in the city.

The work is identified by Gilbert and Phillipson as a 'sonic rebellion against the hegemonic exercise of colonial power [which] involved an assertion of cultural resilience that often ended with the statement, "we are still here"' (Gilbert and Phillipson 2015, 10). I want to continue this analysis by suggesting that it does something to the ecology of the city, past, present and future by acknowledging colonialism as the root of climate crisis. By calling ancestors through sonic graffiti and the phrase 'we are still here' there is an implication that Indigenous people will continue to be here in the future. The sonic tagging marks the city sites with Indigenous presence and resists a history of forgetting. The whispered words 'we are still here' are also reassurance and solidarity to ancestors who suffered from the British imperialist project (such Pocahontas). The tagging provokes questions that interrogate what it means to acknowledge, name and properly reconcile the violence of colonialism in the United Kingdom. There has been no truth and reconciliation process undertaken in the United Kingdom. Writer and legal scholar Kojo Koram called for one in a 2020 article for *The Guardian* newspaper, saying 'The prevailing attitude we hold to our imperial history is one of ignorance and complacency: the British empire is something that we should be proud of, but not pay any great attention to. It should be admired but remain unexamined – the mood music of the British psyche, rather than a cause of current events' (Koram 2020). Morin's sonic graffiti plays in reaction to this mood music, calling it to the foreground. Koram goes on to call for a Truth and Reconciliation process, that asks broad questions about how the United Kingdom has profited from Empire, which is needed if the United Kingdom is ever going to meaningfully grapple with its systemic inequalities. Morin's work addresses the ignorance and complacency of the United Kingdom in ongoing colonialist effects in Indigenous terms and through Tahltan cultural practices.

Marking Indigenous historical, current and future presence in the city through art offers potential for reshaping harmful colonial systems which underpin both social inequalities and climate crisis. Morin's performance centres Indigenous worldviews. Algonquin theatremaker Yvette Nolan writes of Indigenous theatre being used to make medicine, which builds community, negotiates solidarity and performs acts of remembering (2015, 3). Morin's work builds community with both ancestors and the more-than-human actants in urban spaces while performing acts of remembering. Morin makes these links when reflecting on his cultural graffiti intervention at the Tower of London, he says:

> The tower of London was a dream. Understanding how the crown jewels and the monarchy influence Indigenous ways of knowing, it's all a dream until you stand, actually stand there and sing.... and singing the song is also acknowledging this bloody history, and also resource extraction, this creation of this place.
>
> *(Morin quoted in Gilbert and Phillipson 2015, 10)*

Morin makes connections between 'bloody' history and resource extraction to the suppression of Indigenous ways of knowing. The colonial project and ideology of empire has ongoing effects in what is now known as Canada and Tāłtān Konelīne (Tahltan) territory in northern British Columbia. Colonial ideology has shaped Indigenous experience, marginalisation and violence from the state in Canada, including suppression of Indigenous rights, the violence of residential schools and management and use of lands and waterways (Coulthard 2014). The Crown Jewels in the Tower of London are materially connected, through violence and extraction to Indigenous experience in Canada (and other 'commonwealth nations' who endured British colonialism). The places that Morin 'tags' in the colonial metropolis of London are imbued with agentic cultural graffiti that marks the ongoing colonial project, resisting colonialism and hopefully reshaping relationships between people and land in light of climate crisis.

London is a haunted landscape. It is haunted by the colonial and ecological violence that built the city, restyled in a neo-liberal fashion of a modern, cosmopolitan city. Tsing et al. conceptualise the haunting of the Anthropocene, writing 'the winds of the Anthropocene carry ghosts – the vestiges and signs of past ways of life still charged in the present...Our ghosts are the traces of the more-than-human histories through which ecologies are made and unmade' (Tsing et al. 2017, 1). *Cultural Graffiti* calls forth those traces of more-than-human histories of the landscapes and monuments that hold colonial vestiges and residues. Morin wanted the stones that hold up this colonial empire to know that Indigenous people are still here, these songs were also reminders that these ancient relationships are important. The wilful or enforced 'forgetting' of the violence of colonialism shapes London as a landscape, including its moments and tourist sites: 'Forgetting, in itself, remakes landscapes, as we privilege some assemblages over others. Yet ghosts remind us. Ghosts point to our forgetting, showing us how living landscapes reimbued with earlier tracks and traces' (Tsing et al. 2017, 6). Resisting the pressure to forget, this work engages in a remembering of the colonial and ecological violence that lies as an invisible layer blanketing the city. Morin's sonic tags call forth these human and more-than-human ghosts with the words 'we are still here'.

While Morin's work is activist gestures through Tahltan artistic practices, Fast Familiar presents a very different type of work that draws on community to consider climate policy.

The Strategy Room

2023. I am seated around a rectangular table in a blackbox theatre at the University of Reading (UK) amongst an audience of university colleagues and students. There is a data tablet in front of each of us. Our performance is facilitated by Rachel Briscoe of Fast Familiar. In an opening animation, we hear:

> 80% of people in the UK are concerned about climate change. We know it's a big problem, but on our own, it's hard to know what to do about it, especially when there are so many other things to worry about. Right now the UK isn't on track to meet its targets on climate change. That's bad news for everyone. We need a new plan… So you're here today to decide on a strategy for how to get there.
>
> *(Fast Familiar 2023a)*

Over the next 90 minutes, we are be tasked with deliberating over policies to help cities reach net-zero carbon targets. We cover Food and Transport strategies. We watch introductory animations contextualising the problems in Food and Transport before considering some of the proposed interventions. We hear 'postcards from the future', which are performers on video from a not-too-distant 'future' reporting back on how the policies are working in practice including the challenges and advantages to each. Throughout, we rate and then vote on the strategies based on whether we would recommend them for our local area. This anonymised data was then sent to local governments to help inform net-zero policies.

The Strategy Room uses interactive performance methods to create an engaging, democratic and pleasurable experience for participants. Fast Familiar (led by Rachel Briscoe, Dan Barnard and Joe McAlister) are a group of theatremakers and artists from different creative and academic backgrounds who design 'audience-centric' experiences frequently using digital technology. Their work is often interactive and participatory, driven by the audience and democratic principles of discussion, rating and voting, usually using screens and tablets.

The Strategy Room was commissioned by Nesta as a public engagement project on net-zero policies. Part of Nesta's mission is to 'create a sustainable future where the economy works for both people and the planet' (Nesta 2023). They describe *The Strategy Room* as 'an immersive experience that combines facilitated deliberation, interactive polling and collective intelligence to identify the climate change policies that will best help local areas to reach net zero emissions' (Nesta 2023). However, Nesta's language of a sustainable future based on a net-zero economy is mired in a neoliberalist agenda. Brazilian Indigenous activist Ailton Krenak calls sustainability a myth: 'No company on this earth is sustainable, no matter what they say. But these companies and even whole sectors opportunistically appropriate the concept of sustainability for its marketing value' (Krenak 2020, 21). The language of net zero is also misleading. It is not really zero emissions and may hide carbon-intensive practices through a system of 'offsets'. According to Fast Familiar, it is also not an inspiring concept that is easily understood by general audiences. They therefore made the choice not to use the language of 'net zero' in the performance itself (Barnard and Briscoe 2024).

The performances in the 'postcards from the future' featured in the experience were an innovative dramaturgical tool allowing an audience to picture and understand what these policies would look like in practice and what their effect would be. These testimonies were played at specific moments and provided further context to the discussions around the table. They were from three carefully selected characters that represented different geographies, viewpoints, identities and values. We would follow the journey of the character from when the policy was announced to then later seeing it in action and understanding the effect on their daily lives, including any challenges. The characters spoke about the policies in a personal way, detailing how they affected them and their social circles. These performances from the future helped the audience to envision and imagine what these policies could look like in their everyday lives in the not-too-distant future. In my performance, we referred to them often as well as trying to map them on to our local city context. Who did the policies work for? Who did they disadvantage? What might the unintended effects be? Would they work in our specific area? Fast Familiar extensively tested the performance experience in draft form and were able to record potential oppositions to the policies and then write responses to those into the show.

In my performance, we watched a video postcard from Stella who lives in a big city in the not-too-distant future describing how the public transportation subscription service worked in her area. She described how she used a monthly subscription app that provides all the options for travel – other than an individually owned car. You input where you want to go and it gives you the quickest options based on traffic and weather including electric buses, trains, the tube, e-scooters, e-bikes, different types of bikes like cargo bikes for putting your shopping in, e-mopeds, e-cars, hail-a-ride cars like Uber. There are also accessible travel options as well. She described the ease of the app for city living, allowing people to have an array of options at their fingertips with extra incentives for the lowest carbon options like bikes. It allowed her to cycle more than she did when she had a bike collecting dust in her garden shed. She also talked about the disadvantages of the scheme saying she didn't ever cycle in the rain so had to take different options when that happened. Overall, Stella tells us that it is easier and cheaper than private ownership of a car. Her testimony was then used to help us as audiences discuss, rate the policy and decide if we would recommend it in our area. The audience in my performance felt it was a good idea and could work well in Reading (Figure 11.1).

FIGURE 11.1 Image from Fast Familiar's *The Strategy Room*, 2023.

The performance asks audiences the question: What if life could be better? Better for you, better for your community and better for the ecological environment? Hope and optimism for the potential to enact different ways of living marked the performance. Geographer Lesley Head characterises this as slightly out of touch: 'There is deep cultural pressure in the West not to be "a doom and gloom merchant". Hence, even when the evidence points towards the strong possibility of some catastrophic scenarios, the tendency is to focus on policy and action on the most optimistic end of the spectrum of possibilities' (Head 2016, 2). She separates hope from naïve optimism, instead looking to action to generate hope. Rather than doom and gloom, the audience of *The Strategy Room* left feeling hopeful because of the opportunities they had for action in the performance: there are policies out there that could move the needle on climate and would improve our daily lives as well. All we need to do is enact them, which is of course much easier to say than do. The policies are on the more optimistic end, but they are grounded in current realities of policies that are being enacted, trialled or considered in the United Kingdom or Europe. It is probably unlikely that all of these policies will be enacted as quickly as needed, and therefore they may not be able to prevent catastrophic futures. However, they offer a way of envisioning a different way of living together as a first step. This is echoed by an audience member named June, an older woman participant from the performance in Sandwell:

> For me, it was just "the ice caps are melting" and "well, there is nothing we can do about it". But with you explaining the things that can slow it down, it is a positive insight into what the future would be like.
>
> *(Fast Familiar 2023b)*

The work mobilises hope through the examples from the postcards from the future, understanding how everyday people's lives are being affected by the policies. It was also mobilised through the sense of action of voting and the opportunity to discuss the policies with fellow audience members with different perspectives and experiences, who are also local community members.

In her blog detailing reflections from the project, Briscoe writes that

> we need more positive visions of the future. We really don't need any more dystopias or disaster scenarios - existing films, TV shows and books have those covered. We need to help people imagine what a positive future could be like, because, put simply, until we can imagine it, we can't make it happen.
> *(Briscoe 2023)*

Fast Familiar drew on research from the UCL Climate Action Unit about the importance of action in climate storytelling. They felt that the public was ahead of policy makers in some ways as they were more willing to embrace the net-zero policies. They were interested in what narratives would convince policy makers to act (Barnard and Briscoe 2024). Researchers De Meyer et al. argue that 'climate change is primarily conceptualized as a threat we should be concerned about, rather than as something we know how to act on' (De Meyer et al. 2021, 1). This leads to a 'poverty of stories' because the existing narratives around climate change tend to use the same language and focus on raising awareness and concern. Consequently, this then leads to a perceived 'lack of agency', or inability to act, among the public. This creates a sense of only two limited options for action: consumer choices and climate activism. The authors argue that moving from issue-based stories to action-based stories would provide a sense of agency, as there is 'evidence from psychology and neuroscience showing that "in real life actions drive beliefs"—rather than awareness, beliefs or concern leading to behavior change' (De Meyer et al. 2021, 2). The democratic approach to the performance felt empowering and gave the audience a sense of agency through visible action.

Both of these performances make interventions into the way we live in cities in very different ways. As urban ecologies become essential in a time of increased urbanisation and climate crisis, we need cultural acts of imagining different ways of living together in more just and equitable ways. While Morin's work embodies Indigenous knowledge and culture, *The Strategy Room* focusses on governmental policies at local levels. I suggest that the combination of principles from each work allows us to see the potential of performance in imagining a more climate just future. Taking the colonial marked systems in the United Kingdom and transforming them into community-driven ones that prioritise people and planet is where we need to begin. The Indigenous agency imbued throughout the city gives us a knowledge system to base future strategies on. As Morin experienced at the Tower of London, what would it mean to acknowledge and then reshape the legacy of Empire that is based on extractivism and exploitation? The hopeful community-driven policies of *The Strategy Room*, injected with Indigenous ways of knowing may be part of the answer.

References

Barnard, D. and Rachel Briscoe. 2024. 'Interview with the Author.' London, 29 October 2024.

Briscoe, Rachel. 2023. 'The Strategy Room 1/3: Creating a Design Brief.' *Fast Familiar*, 8 May. https://workroom.fastfamiliar.com/the-strategy-room-design-brief, accessed 20 October 2023.

Coulthard, Glen. 2014. '#IdleNoMore in Historical Context.' In The Kino-Nda-Niimi Collective, ed, *The Winter We Danced: Voices from the Past, the Future, and the Idle No More Movement*. Winnipeg: Arp Books, 32–37.

De Meyer, Kris, Emily Coren, Mark McCaffrey, and Cheryl Slean. 2021. 'Transforming the Stories We Tell about Climate Change: From "Issue" to "Action".' *Environmental Research Letters* 16 (1), Article 015002. https://doi.org/10.1088/1748-9326/abcd5a.

Dorries, Heather, 2023. 'Indigenous Urbanism as an Analytic: Towards Indigenous Urban Theory'. *International Journal of Urban and Regional Research* 47 (1), 110–118. https://doi.org/10.1111/1468-2427.13129

Fast Familiar. 2023a. *The Strategy Room*. Reading.

Fast Familiar. 2023b. *The Strategy Room – Documentation Video*. https://youtu.be/i8lIVDwRRfY?si=MsepjsXkwse7_vCv, accessed 20 October 2023.

Gaston, Kevin J., ed. 2010. *Urban Ecology*. Cambridge: Cambridge University Press.

Gilbert, Helen and J. D. Phillipson. 2015. 'Cultural Graffiti in London: Singing Life into Exhibitions and Embodying the Digital Document.' *UNESCO Observatory Multi-Disciplinary Journal in the Arts* 5 (1), 1–36.

Head, Lesley. 2016. *Hope and Grief in the Anthropocene: Re-Conceptualising Human-Nature Relations*. London: Routledge.

Huarcaya, Sergio and Dani Phillipson. 2013. *The Artist Sings: Peter Morin in Conversation*. https://vimeo.com/119944337, accessed 9 September 2023.

Koram, Kojo. 2020. 'Britain Needs a Truth and Reconciliation Commission, Not Another Racism Inquiry.' *The Guardian*, 16 June. https://www.theguardian.com/commentisfree/2020/jun/16/britain-truth-reconciliation-commission-racism-imperial, accessed 20 October 2023.

Krenak, Ailtron. 2020. *Ideas to Postpone the End of the World*. Translated by A. Doyle. Toronto, ON: Anansi International.

Nesta. 2023. 'The Strategy Room.' https://www.nesta.org.uk/project/strategyroom, accessed 10 October 2023.

Nolan, Yvette. 2015. *Medicine Shows: Indigenous Performance Culture*. Toronto: Playwrights Canada Press.

Robinson, Dylan. 2019. 'Speaking to Water, Singing to Stone: Peter Morin, Rebecca Belmore, and the Ontologies of Indigenous Modernity.' In Victoria Lindsay Levine and

Dylan Robinson, eds, *Music and Modernity among First Peoples of North America.* Middletown, CT: Wesleyan University Press, 220–39.
Thrush, Coll-Peter, 2016. *Indigenous London: Native Travelers at the Heart of Empire.* New Haven: Yale University Press.
Tsing, Anna, Nils Bubandt, Elaine Gan, and Heather Anne Swanson, eds. 2017. *Arts of Living on a Damaged Planet.* Minneapolis: University of Minnesota Press.
Woynarski, Lisa, 2025. Performing Urban Ecologies. Cambridge: Cambridge University Press.

12

MUTATING PERFORMANCE

On Monira Al Qadiri's Speculative Petroaesthetics

Katia Arfara

In performances, sculptures, installations and films, Monira Al Qadiri's work centres on the ruinous effects of the rapid social and economic change from pearl to petroleum culture in the Arabian Peninsula and the oil extraction that destabilises ecosystems and saps the planet's resources. By interweaving the maritime world of pearl diving with oil's material manifestations, the artist engages with the unsustainability of carbon-intensive fossil fuel production and consumption and the regulative heteronormativity of the petrocapitalist economies in the Gulf. In this chapter, I focus on Al Qadiri's immersive video installation *Diver*, a four-minute choreography of four synchronised swimmers performing in the open sea off Abu Dhabi at night.[1] *Diver* premiered at Warehouse 421 in 2018 in Abu Dhabi and was commissioned by Durub Al Tawaya (DAT), Abu Dhabi Art's performing arts program, and the Asia Pacific Triennial of Contemporary Art (APT). I analyse *Diver* as as a key work in Al Qadiri's ongoing critical engagement with petroaesthetics in the Gulf; it is an exemplary speculative reflection on the re-historicisation and feminisation of the pre-oil and post-oil eras. Here, 'petroaesthetics' implies the alien and alienating effect of oil as a commodity in the global economy and politics, but also in the twenty-first-century cultural imaginary.

I argue that *Diver* is the culmination of Al Qadiri's interactions with the material public expression of petroaesthetics and nation-making cultural traditions. She interweaves historical memory and oil modernity in an uncanny watery environment of swimmers and pearl music singers. Instead of representing the oil age in which she lives, Al Qadiri offers glimpses of the world that comes after oil has become obsolete as an energy source. She makes visible the usually invisible ways in which oil shapes representation in architecture and infrastructure, icons

DOI: 10.4324/9781003584445-16

and symbols, images and artefacts to investigate how pre-oil heritage mutated into national and cultural formations. In order to place the entanglement of oil and gender in *Diver* into context, I analyse these themes in a few of Al Qadiri's early photography and video performances, then I examine how *Diver* deepens her ongoing investigation of gender performance as a way to unsettle patriarchy, petroaesthetics and representations of 'authentic' cultural heritage, with a view to a world after oil.

Born in Senegal, Al Qadiri grew up female in Kuwait's male-dominated society and belongs to a generation of artists who came of age during the young nation-state's rapid transformation. During her lifetime, the oil industry, which had grown exponentially since the exportation of the first barrels of oil in 1946, transformed her homeland from a vibrant commercial settlement and regional entrepot in the northwest corner of the Gulf into a global geopolitical player (Al-Nakib 2016, 91–94). Al Qadiri first encountered the black liquid when Saddam Hussein's forces set hundreds of oil wells on fire as they retreated from Kuwait to Baghdad during the 1991 Gulf War. While the oil infrastructure largely existed in heavily restricted areas in Kuwait as a metonymy for limitless growth, the wells made oil visible and tangible in all its toxicity when they burned for months on end, polluting the sky, the sea, the soil and every aspect of Kuwaitis' everyday lives in what is today considered one of the largest environmental disasters of the twentieth century (Al-Nakib 2023, 28–33). The oil fires highlighted the fragility of petrocapital accumulation and nationhood. The war and the unprecedented environmental catastrophe affected not only the artist's perception of Kuwait's economy but also her identity as a Kuwait-raised woman. During the Iraqi invasion of Kuwait, she stayed at home with her older sister, the performance artist and musician Fatima Al Qadiri, and began cutting her hair very short and crossdressing, as a reaction to the dysfunctionality of gender roles within the Arabian Peninsula (Brodsky and Olin 2012, 74). Photographed by her sister and entitled *Bored 1997*, Monira Al Qadiri staged a series of portraits inside the family house, where she dressed in traditional Kuwaiti menswear (*dishdasha*), Western male clothing, military uniforms and carnival masks while performing small, everyday and unheroic gestures such as combing her short hair, speaking on the phone or laughing in front of a white wall with hunting trophies. The unsettling of masculinity by crossdressing becomes even more subversive with the integration of this playful rendering of stereotypical male social roles in domestic interiors, displacing the male gaze from the public sphere of a nation to the feminised interior of an ordinary home. Monira and Fatima Al Qadiri's *Bored 1997* underlines the value of witnessing war and environmental catastrophe from the margins, showcasing how distant such events can feel to women who are purposely sidelined. This series of portraits, now often exhibited as a large photographic collage measuring 164 by 200 centimetres, constitutes Al Qadiri's first assertion of a mutant identity (she was 14 at the time), which grew into a conscious undoing of rigid societal norms in the Gulf in her subsequent music videos and later in *Diver*.

Building on this first series of staged photographs in her later video performances, Monira Al Qadiri expanded on the idea that aspects of identity and symbols of petromodernity are interrelated. In these music videos, she often plays the role of male singers and shows images that connect to the lyrics and her own questioning of the entanglement between oil economies and gendered nationalism. She appropriates this pop medium to create alienating self-portraits such as *Torment* (Wa Waila, 2008), one of her earliest video works on gender in male-driven Kuwaiti society, filmed when she was living in Tokyo and completing her PhD on the aesthetics of sadness in the Middle East. It represents a Kuwaiti folk song by Abdul Wahab Al Rashid about lost love, displacement, gender and death in a poetic and uncanny way. She films herself as the male singer while a chorus of male dancers dressed as women in long black wigs and black dresses. Swinging their long 'feminine' hair, the chorus imitates folkloric dances that women in the Gulf usually perform on social occasions. The camera follows this unorthodox chorus to various locations such as a *majlis* (sitting room), the seashore, abandoned mud-brick houses and a cemetery. Another video performance, *Father of Pain* (Abu Athiyya, 2013), is based on an Iraqi lamentation song (*mawwal*) entitled 'I Spent Years of My Life with You'. In a dark, funereal setting, Al Qadiri plays the famous Iraqi singer Yas Khodhor and performs a dance with daggers, interweaving Khodhor's actual voice with a traditional popular dance that has deep roots with the Kawliya people, a marginalised Romani group in Iraq. In both these early video performance works, *Oh Torment* and *Father of Pain*, Al Qadiri complexifies gender representation. She plays a male singer of pre-oil Kuwaiti or Iraqi folk songs, but many aspects of her performance are 'mismatching' and somehow 'wrong'. As she says, 'the costumes are wrong, the makeup is wrong, the beard looks fake [...] but it works' (Al Qadiri and Alsaden 2021). The apparent fakeness of the costumes and Al Qadiri's beard—for her, a symbol of machismo—interrogates the entanglement of masculinity and the oil economies, producing uncanny feelings because everything is not in quite the 'right' place. However, she does not erase her female characteristics; Al Qadiri unmakes masculinity from the outside, preventing the representation of a male-driven society from consolidating itself as an impenetrable body. By crossdressing in her video performances, the artist attempts to unsettle the Gulf's normative petrocultures, where in order to have power and agency, one must be male and directly connected to oil-centric, environmentally destructive institutions. In Al Qadiri's mutant world, everything is porous, infinitely negotiated and reconfigured.

With *Diver*, Al Qadiri continues to critically explore this profound cultural and social transformation in the nomadic communities of the Peninsula, which have been significantly impacted by the industrialisation of oil. *Diver* features

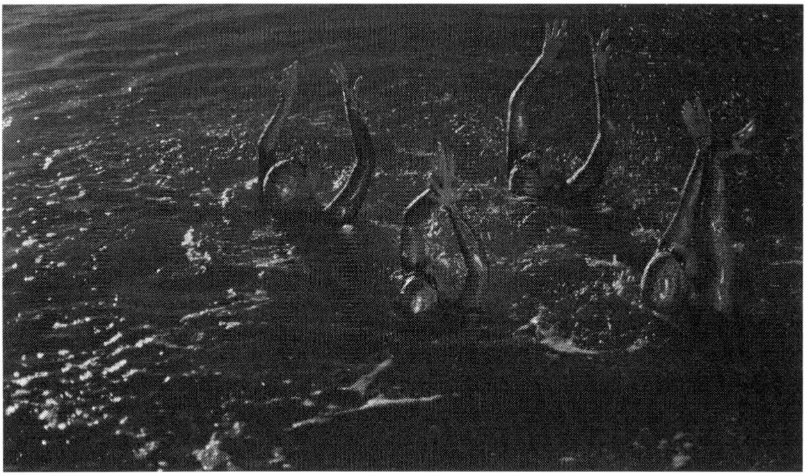

FIGURE 12.1 Monira Al Qadiri, still from *Diver*, video installation, 2018. Image courtesy of Monira Al Qadiri.

both pre-oil pearl diving songs and the decidedly post-oil phenomenon of seemingly oil-slicked water. The swimmers wear goggles and iridescent head-to-toe bodysuits that resemble the sheen of both pearls and oil, and the ocean in which they swim appears black (see Figure 12.1). The singing is accompanied by the jarring, dissonant sound of oil pipes knocking together.

The camera follows the movements of four professional synchronised swimmers choreographed to a Bahri seafaring folk song from the early 1960s, before the genre became part of Kuwait's national heritage and entertainment industry. Fidjeri music was performed on ships during long pearl diving journeys, but also on land between trips across the Gulf region, assimilating various cultural influences spanning the Gulf and Africa. Fidjeri is mainly vocal (a series of sections sung by a soloist and a male chorus) but also involves dance movements. It is performed only by men—like the artist's paternal grandfather Issa Al Qadiri—and the voices are accompanied by syncopated clapping and drumming (Touma 2003, 91–92). Verses of these work songs describe harsh living and labour conditions and often contain prayers and spiritual rituals. The lyrics of the song that Al Qadiri incorporates translate to:

> Ah. Ah. Oh, world. Time has come down on me like a massive weight. Rounded up its pals like an army and ganged up on me. Damn you, world. You've disappointed me. You and your gang. You treat the bastard better than you do me. But my buddies here are like flowers, and even better.
>
> *(Al Qadiri 2015, 40)*

These songs were intimately connected to diving for pearls, the only local commodity with an international market in the pre-oil era. The industry remained in the hands of local entrepreneurs who 'operated in symbiosis with rulers', which led to pre-oil urbanisation in the second half of the nineteenth century (Fuccaro 2001, 176). The pearl industry in the Gulf flourished until the early 1920s, but it started to slowly decline in 1929, partly because of the Depression and partly because of competition from less expensive Japanese cultivated pearls. The discovery of oil catalysed the end of pearl diving. The emergence of the oil industry transformed traditional industries and trade, as well as the premodern urban environment, by incorporating local populations into oil-related labour on temporary construction, pipeline or transport work (Macdonald 2017, 298).

In *Diver*, as in her other performative works and music videos, Al Qadiri places mismatched elements together. She interweaves the male-dominated world of pearl divers with synchronised swimming, a highly disciplined water sport that combines athletics with artistic performance. Thanks to the lighting effects she orchestrates, the swimmers move in water that looks like oil, meant to evoke the many offshore oil fields found in the Gulf and the oil spills that often result from this extraction. The clanking oil pipes weave into the voices, adding another layer to this paradoxical performance. The mismatching of the choreography with the sound environment underlines the historically male-dominated worlds of oil and pearls. At the same time, the video highlights the uncanny location of this synchronised swimming, which normally takes place in protected environments such as swimming pools. Aside from the fidjeri music that functions as a marker of time and place, these dark waters can be localised but not territorialised—they can be located nowhere and anywhere.

The four female swimmers acquire an uncanny appearance, acting as one plural, fluid body—*a* diver. This body seems to swim in oil and dive into the deep, geological time of this natural resource formed over millennia. Filmed close-up or from above, these female bodies transform the entire ocean into an inclusive, feminised space in which the obsolescence of the drill bits encounters the ghostly melancholy of the pearl divers' music. The beauty of the aquatic choreography contrasts with the seeming toxicity of the black waters, while echoing the 'beautification' of fidjeri music for entertainment as a performative enactment of heritage, tradition and authenticity (Andraos 2019, 64–67). The oily-looking ocean and the performing-as-one-body female swimmers encounter the male singing chorus (which the viewer never sees) and become a mutant creature that questions patriarchal representations of the Gulf's pre- and post-oil cultures (Al Qadiri 2023, 113). The natural history of pearls found in the deep sea interlinks with the anthropocentric history of oil, both part of the same extractive logic. *Diver* complexifies the normative, homogeneous landscape of the fossil fuel economy and reflects its alienating condition.

I argue that *Diver* is the performative manifestation of Al Qadiri's ongoing project, *Alien Technology*, a series of public sculptures that take the drill head as their main motif. In *Alien Technology*, as in *Diver*, Al Qadiri mingles this oil industry technology with the seafaring world of pearl diving, utilising fibreglass, a petroleum-based material coated in dark, iridescent automotive paint, to make these speculative connections. The uncanny iridescence of *Alien Technology* mimics the dichroic colour of both oil and pearls and stands for multiplicity because it appears to contain several colours simultaneously. Dichroism has a non-normative, liminal appearance in which nature and technology coexist as the two limits that provide for a paradoxical unity. Pearls and oil come together as solid components of a speculative transnational history of the pre- and post-oil eras. Both of these periods are part of Al Qadiri's biography as a child of the post-oil generation and the granddaughter of a fidjeri musician.

In Al Qadiri's *Diver*, oil, drills and pearls merge in an 'alien choreography' and the exhausting physical action of the four swimmers mirrors the labour of the pearl divers. Appearing and disappearing in the dark water off Abu Dhabi, their collective body forms a mutant creature that simultaneously reflects the colour of pearls and oil, constantly changing and capturing light while interweaving with the fidjeri song and the distant sound of the drill in the background.

Diver's choreography mimics the function of a drill head; the four swimmers could be perceived as the four moving parts of the oil drill, performing abstract drilling movements towards the sky (see Figure 12.2). The darkness of the water and the reflections off the swimmers' iridescent bodysuits create an uncanny

FIGURE 12.2 Monira Al Qadiri, still from *Diver*, video installation, 2018. Image courtesy of Monira Al Qadiri.

watery environment that is constantly mutating. Like in the *Alien Technology* sculptures, the swimmers turn the violent process of extraction upward, reversing their initial earthward direction. Al Qadiri feminises the phallic nature of the drill head through the bodies of the swimmers by contrasting the flexibility and grace of the female body with the rigid materiality of the drill (Al Qadiri 2015, 38). In both sculptural and performative ways, Al Qadiri unsettles the functionality of the drill heads and defamiliarises their appearance, contributing to the object's transformation into a biomorphic 'thing' that is not easily recognisable as a drill head.

Al Qadiri imagines herself as embodied by this plural female body of four swimmers submerged in and re-emerging from the dark oil-like water in an ongoing movement that only ends when the song does; it is a mysterious spectacle with no visible audience performed in the dark as a poetic embodiment of a dystopic, post-oil future. The dark liquid stands for the plurality of the self, an alien lifeform in constant mutation that escorts us from the past to the future. In *Diver*, oil is a powerful narrator: the camera seems to be moved by oil, set in motion by its own subject. Oil is the only nonhuman element that can entangle regional-turned-national formations with contemporary petrocapitalism.

I contend that Monira Al Qadiri's work has some parallels to philosopher Reza Negarestani's semi-fictional book *Cyclonopedia* (2008), which tells the nonhuman story of oil. This work of theory-fiction is based on the 'field notes' of a famed Iranian archaeologist who claims that oil is a terrible Tellurian entity with a subterranean logic of its own. 'The cartography of oil as an omnipresent entity [...] is the undercurrent of all narrations, not only the political but also that of the ethics of life on earth' (Negarestani 2008, 19). The oil becomes 'the narrative organiser', as 'there is no darkness in this world which has not its mirror image in oil' (Negarestani 2008, 19). Negarestani speculates on the new world that might come after this short 'petroleum interval', where macho-exceptionality and normative heteropatriarchy are not abolished but are held responsible for the destruction of the planet (Negarestani 2020, 131). Building on Negarestani's call (2020) for new cartographies that mend the split between humanity and nature, I argue that in Al Qadiri's *Diver*, the oil-like water, the costumes and the mismatched soundscape combine to produce a fluid space, where the human and nonhuman, material and immaterial coexist in a speculative experience of the present.

Using forms, materials and colour scales related to the oil industry, Al Qadiri is not simply interested in a counter-history of the region but rather in incorporating the effects of fossil fuel extraction into the future, inviting the audience to flow from the past backward into the future and to think about the true price of wealth and prosperity in the Gulf. Her early music videos draw from her experience growing up female in a patriarchal Kuwaiti society, and *Diver* also alludes to her claim of a fluid identity. It is part of her ongoing self-portrait, whether she is directly using her face and body as she did in *Oh Torment* and *Father of All Pain*, or whether she turns her lens towards an abstract narrative such as *Diver*'s watery choreography of a femininised oil drill.

In *Diver*, oil becomes a mutant element that connects all events on Earth as it inserts itself into every orifice of our modern existence, even if at times it passes entirely unnoticed, 'an ancient alien' who lives in our midst (Al Qadiri 2023, 113). Oil, this chameleon-like character, both a curse and a blessing, is an other-than-human storyteller who holds together Al Qadiri's dramaturgy of possible future worlds, where all kinds of life would be accounted for and concretely accommodated. For both Al Qadiri and Negarestani, imagining new worlds means imagining new lifeforms that are not simply dying but are constantly in flux, determined by what Negarestani calls 'rational imagination', or a way of worldmaking that holds onto 'the responsibility to account for what the world should look like despite how the current world is' (2020, 132–33). Visualising the transformative nature of oil and its often-invisible effects on our everyday life through these four female dancers, Al Qadiri interrogates the entanglement of masculinity with pearl culture and fossil fuels—in the Gulf, petrosocieties need the past to legitimise their continuity and futurity. The mismatching of the watery choreography with the pearl divers' music de-essentialises the pre-oil 'heritage' symbols of pearl diving while reflecting on a feminist-driven future after oil.

By privileging fluidity over fixity in *Diver*, as well as in the early video works I examined here, Al Qadiri unsettles the region's hegemonic discourse of cultural exceptionalism and its essentialisation of tradition in the service of nation-building. Because the drill is a phallic symbol of petromodern expansion, it materialises the patriarchal structures of the petroeconomy in which the artist was raised. Through the synchronised swimming, Al Qadiri beautifies the oil drill heads and makes them seductive in the way that oil wealth is seductive. However, she is not interested in looking at them from the perspective of the present; she prefers to see them from a (dark) future as a reminder of a once destructive and now useless way of living. In a narrative that shifts between troubled questioning about what is coming next and sombre retrospection on a world already passed, Al Qadiri creates a speculative space, reflecting on our future without abundant oil.

Diver straddles pre- and post-oil life in the Gulf, entangling the fluid maritime realm of seafaring and pearl diving with the alienating effect of the oil industry. By interweaving symbols of national heritage such as fidjeri music with those of drill-modernity, the artist defies Kuwait's self-representation as an ahistorical and future-forward nation, fostering (historical) memory as a way to imagine the Gulf *after* oil. Al Qadiri's video performance could be perceived as a performative counter-monument not only to petroaesthetics but also to the pearl diving tradition. *Diver* emerges from the 'sea' at a moment of impending ecological catastrophe as a warning about the unsustainability of carbon-intensive industry but also of the rigid societal structures on which the Gulf petrostates have been built. Al Qadiri speculates on a woman-dominated age after oil, when petroleum and patriarchy could be part of the Gulf's 'heritage' and mutation will live on in hybrid and heterogeneous forms of national and cultural histories.

Al Qadiri's *Diver*, as some of her earlier work, introduces more-than-human representations where the relations between ideas and forms, shapes and colours, human and the Earth, gender and identity, communities and the planet are constantly performed. She imagines herself in a constant flux, refusing to belong to one place or one narrative. Mutation is here perceived not as a new ontological understanding of the body, but as inherently performative, a metonymic register that stands in for the hybridity of culture, gender and identity.

Note

1 Costumes: Jasmina Popov-Locke; Choreography: Isabelle Tan; Dancers: Isabelle Tan, Carly Athawes, Zsofia Péres, Daria Galkina.

References

Al-Nakib, Farah. 2016. *Kuwait Transformed: A History of Oil and Urban Life*. Palo Alto, CA: Stanford University Press.

Al-Nakib, Farah. 2023. '"No Words around to Describe": Between Seeing and Comprehending Kuwaiti's Oil Fires.' In Nelida Fuccaro and Mandana Limbert, eds, *Life Worlds of Middle Eastern Oil: Histories and Ethnographies of Black Gold*. Edinburgh: Edinburgh University Press, 28–52.

Al Qadiri, Monira. 2015. 'Choreography with Alien Technology.' In Sophia Al-Maria, ed., *The Happy Hypocrite – Fresh Hell No. 8*, London: Book Works, 38–41.

Al Qadiri, Monira. 2023. 'Dark Genesis.' In Thomas D. Trummer, ed., *Monira Al Qadiri: Mutant Passages*. Bregenz: Kunsthaus Bregenz, 110–13.

Al Qadiri, Monira and Amin Alsaden. 2021. 'Monira Al Qadiri Dives Deep into Oil.' *Ocula*, March. https://agb.life/wp-content/uploads/2021/03/Monira-Al-Qadiri-Ocula-Conversation.pdf, accessed 17 April 2025.

Andraos, Amale. 2019. 'Problematizing a Regional Context.' In Harvey Molotch and Davide Ponzini, eds, *The New Arab Urban: Gulf Cities of Wealth, Ambition, and Distress*. New York: New York University Press, 58–78.

Brodsky, Judith K. and Ferris Olin, eds. 2012. *The Fertile Crescent: Gender, Art, and Society*. New Brunswick, NJ: Rutgers University Institute for Women and Art.

Fuccaro, Nelida. 2001. 'Visions of the City: Urban Studies on the Gulf.' *Middle East Studies Association Bulletin* 35 (2), 175–87. https://doi.org/10.2307/23063464.

Macdonald, Graeme. 2017. '"Monstrous Transformer": Petrofiction and World Literature.' *Journal of Postcolonial Writing* 53 (3), 289–302. https://doi.org/10.1080/17449855.2017.1337680.

Negarestani, Reza. 2008. *Cyclonopedia: Complicity with Anonymous Materials*. Melbourne: re.press.
Negarestani, Reza. 2020. 'On an Impending Eternal Turmoil in Human Thought.' In Suad Garayeva-Maleki and Heike Munder, eds, *Potential Worlds: Planetary Memories and Eco-Fictions*. Zurich: Scheidegger & Spiess, 125–34.
Touma, Habib Hassan. 2003. *The Music of the Arabs*. Portland, OR: Amadeus Press.

Climate Risk, Activism and Eco-Performance

13

OKADA TOSHIKI'S POSTVISUAL DRAMATURGY IN TIMES OF DURATIONAL CATASTROPHES

Kyoko Iwaki

In the year 1212, Kamo no Chōmei, a Buddhist monk and author, completed his draft of *Hōjōki* that begins with the evocative line: 'On flows the river ceaselessly, nor does its water ever stay the same. The bubbles that float upon its pools now disappear, now form anew, but never ending long' (Chōmei 2013, 5). This opening passage, with its quiet yet profound rhythm, starkly contrasts with the subsequent narrative—an almost graphic account of the chaotic disasters that punctuated Chōmei's life. Yet beneath the strikingly vivid reportage, the reader can also feel there lies an unwavering Buddhist sensibility of *mujō*, the empty sense of transience towards life, threading through each calamity like a contemplative undercurrent. *Hōjōki* is often heralded as Japan's 'oldest catastrophe literature', offering a vivid tableau of the five monumental crises that Chōmei personally encountered during the younger years of his life (Kobayashi 2013, 14). The title, commonly translated as *The Ten-Foot-Square Hut*, evokes the small, austere refuge where Chōmei, having forsaken the opulence of his lineage as the heir to Kyoto's Shimogamo Shrine, sought comfort in solitude. It is within this confined space, this metaphorical microcosm of existence, that Chōmei's reflections on the impermanence of life and the inexorable march of time are poignantly articulated.

Having already endured the sorrow of losing his father at the tender age of 18, Chōmei's subsequent decades unfolded amidst a series of relentless calamities. In his twenties and thirties, he faced the Great Fire of Angen in 1177, the Tornado of Jishō in 1180, the abrupt and transient relocation of the capital to Fukuhara within the same year, the Great Famine of Youwa in 1181 and the Great Earthquake of Genryaku, which was followed by a devastating tsunami in 1185 (Asami 2012, 13). Compounding these natural disasters were the incessant civil

DOI: 10.4324/9781003584445-18

wars that ravaged the country, with the Taira and Minamoto clans waging fierce battles for political dominance in the fragmented pre-nation landscape (Asami 2012, 13). Amidst these adversities of fire, tornado, relocation of the capital, famine, earthquake, tsunami and civil wars in less than a decade, Chōmei came to a realisation towards the end of his life that what appears stable, visible and tangible from a human perspective is but an ephemeral illusion—mere flows and bubbles on the surface of a vast, ever-changing river that can vanish in an instant. For Chōmei, the essence of human existence was not defined by states of being, persisting or possessing, but by processes of becoming, withering, vanishing, decaying and fading in concert with the world.

Transposing this existential outlook to Performance Studies resonates with Cary Lavery's ecocritical dramaturgy, which advocates to 'come to terms with weakness and inadequacy' of our livelihoods rather than glorifying human dominion over the force of nature (Lavery 2016, 233). It also aligns with Peter Eckersall's post-anthropocentric concept of 'becoming less', where humanity diminishes 'in relation to the assemblage of forces around us' (Eckersall 2024, 71). By echoing these ideas, Japanese theatre director and playwright Okada Toshiki, alongside his sculptor-collaborator Kaneuji Teppei, envisions a state of what they call becoming 'half-transparent (*hantōmei*)', which acknowledges humans to accept weakness and inadequacies and to exist in this world humbly and harmoniously with those invisible more-than-humans (Iwaki 2021, 163).[1] This view, which underscores the fleeting nature of mortal life, suggests that individuals, while still alive, may be more willing to forge connections with the invisible spirits, disembodied voices and the spectral presences of the dead and the unborn. Such relational complications between the living and the dead—which resonates with many scholars who resist the hegemony of positivism, such as T.J. Demos, Isabelle Stengers and Avery Gordon—offer a pathway to a more entangled, post-human, postvisual dramaturgy, which, owing to its lack of visible, narrative and logical performers, veers towards what could be even deemed a 'non-theatre' (Uchino 2020). Okada and Kaneuji believe that a perspective that sees reality through entanglements of the living/dead, visible/invisible and humans/more-than-humans may more accurately reflect a world perpetually marked by seemingly invisible and non-ending catastrophes (Demos 2013; Gordon 2008; Stengers 2023).

Medieval literature scholar Kobayashi Kazuhiko (2013) recently posited that 'contemporary Japan has entered the era of *Hōjōki*', an assertion grounded in the persistent natural and anthropogenic disasters that plague a nation encircled by the sea, dotted with 33 nuclear reactors (54 before the Fukushima Disaster) and situated on a volatile network of active volcanoes and tectonic plates (29). The central contention of this brief chapter is that, as we go back into the era of *Hōjōki*, with a maddening amount of synchronous and enduring crises erupting both on a mundane and a planetary scale, the dialogues with the supernatural, immortal and invisible beings are becoming more prevalent in the theatres and

FIGURE 13.1 Okada Toshiki and Kaneuji Teppei's *Eraser Forest*, 2020. Photograph by Keizo Kioku. Image courtesy of 21st Century Museum of Contemporary Art, Kanazawa, Japan.

the wider society. In these times of durational ecocritical catastrophes—where the human gaze, or the human capacity for spectatorship, often falters in the face of events that are essentially imperceptible and invisible for humans in time and scale—reawakening the imagination to envision beyond the immediately visible may offer a vital source of solace. When reason only through positivism does not provide either a solution or an answer, a collectively shared imagination that goes beyond the visual could serve as a survival mechanism, enabling humans to look beyond and not look away from planetary calamities (Figure 13.1).

The primary case study for this chapter is Okada and Kaneuji's *Eraser Forest* (*Keshigomu mori*, 2020), a multidisciplinary exhibition and performance that opened at 21st Century Museum of Contemporary Art, Kanazawa. I visited the performance/exhibition on 7 February 2020, right before Japan's initial lockdown for COVID-19. *Eraser Forest* explores the dramaturgy of catastrophe through the metaphor of fading, a poignant commentary set against the backdrop of ominous thoughts towards the approaching pandemic and what Harry Harootunian describes as life in 'permanent parenthesis' following the Fukushima disaster (2015, 29). However, as noted through references to Chōmei's medieval literature, Japan's susceptibility to aquatic, atmospheric and geological hazards is not a rarity but has been a historical constant. When analysing this Japanese contemporary exhibition/performance, therefore, I argue that it is crucial to recognise that living in a state of permanent parenthesis has more or less been a prevailing reality, predating the Tōhoku earthquake, Fukushima nuclear crisis and the COVID-19 epidemic. In this regard, one could say that the Japanese archipelago has perennially inhabited what Eva Horn characterises as a catastrophe that does not merely approach but that 'which is' (2018, 7). Catastrophe, in this sense, becomes a quotidian presence.

In acknowledging the volatile cultural milieu that requests the inhabitants to be in a state of perpetual flux and in-betweenness, I align with Okada in positing that individuals in Japan are engaged in multiple performances of fading. Already on a mundane level, this phenomenon manifests in various forms: from hiding in nature, exemplified by those tiny hermit huts of Kamo no Chōmei; melting into the anonymous mass as those businessmen squashed into commuter train carriages; and the collective conformity observed as individuals blend into groups to avoid provoking others. These manifestations of active fading could be considered, on the one hand, as gestures of peaceful cohabitation, or what can be otherwise described as the Buddhist notion of 'codependint origination' (Rowe 2011). On the other hand, however, it could be interpreted as an expression of post-Fukushima disquietude where people live in a mode of what Harootunian calls as a life in 'permanent parenthesis', avoiding nuisance, evading venture and perennially atrophied in fear of ostracism (2015, 29).

It also resonates with previous scholarship on the 'singular-plural' (Nancy 2000), 'relational' (Glissant 2010) and 'trans-individual' (Cvejić and Vujanović 2023), which in various degrees and modalities suggests a reciprocally damaging yet mutually supportive life with fellow humans, the natural environment and, in Okada and Kaneuji's case, even the deceased. This mode of living through the performance of fading is also a mode of collective survival, deeply ingrained in Japan's catastrophe-ridden cultural soil, reflecting a tried-and-tested strategy for enduring an environment fraught with perpetual uncertainties.

It is crucial, however, not to conflate the action of fading with the somewhat analogous yet distinct practice of hiding. The divergence between these two performances becomes evident when examining their underlying vectors. Hiding often aims at internal protection: it involves concealing one's definitive positionality as a means of surveillance and potential manipulation of others. In contrast, fading is driven by humble intentions: it involves occupying a position of indeterminacy, hesitation and formlessness as an act of queer affect and care (Sedgwick 2003; Puar 2007; Halberstam 2011; Chen 2012). This form of fading entails temporarily suspending a fixed individual stance, dithering and blurring the boundaries between oneself and the unknown, thereby fostering a more 'multi-centric' worldview in solidarity with disenfranchised others (Cheng 2002, xxiii).

Okada Toshiki (b. 1973) is the writer-director of the theatre company chelfitsch, founded in 1997, and is recognised as one of the most perceptive artists responding to the political and artistic challenges posed by the unknown and unforeseen zeitgeist. The triple catastrophe of 11 March 2011 marked a pivotal moment for Okada, as the extensive destruction and the sudden loss of life compelled him to incorporate the voices of the dead into his dramaturgy. In so doing, as Stanca Scholz-Cionca (2021) observes, he pursued artistic possibilities through a 'conscious engagement with the premodern stage art' of Japanese nō plays (19). However, Okada's primary aim in his post-Fukushima works, such as *Current Location* by *Genzaichi* in 2012, *Ground and Floor* (*Jimen to yuka*) in 2013) and *Time's Journey Through a Room* (*Heya ni nagareru jikan notabi*) in 2016 was not to replicate the formalities of nō. Rather, he employs the dramaturgy of the medieval dance play to connect and converse with invisible others, via which potential versions of realities can only be articulated through measures beyond rationality. He expressed a fascination with the 'affinity between ghosts and theatres [as they are both] great inventions of human beings' (Okada 2016, 5–7). In essence, in the wake of the tragic loss of life and the disappearance of vast landscapes and numerous species, Okada sought to utilise the grammar of nō to connect, converse and comingle with those beings existing in the realm of what is called *myō*.

In Buddhist terminology, the land where the invisible, irrational and supernatural reside is known as the world of *myō*. The concept, primarily articulated by Tendai Buddhist monk Jien and revisited by contemporary religious studies scholars such as Sueki Fumihiko, is frequently examined in relation to its counterpart, *ken* (2019, 14–17). Viewed through the lens of *ken*—which aligns with modern human reason and the visible world—*myō* emerges as a space where the unseen, the supernatural and the virtual coexist, serving as a domain for the unknown 'others' (2019). Ultimately, *myō* is a realm that thrives on the collective imagination of human beings, underscoring the close interplay between reality and fiction in sociocultural contexts. When the majority of people believe in the existence of *myō*, as was the case in medieval Japan, it establishes itself in the realm of reality rather than fiction. Conversely, when fewer individuals cultivate the sensitivity to imagine beyond immediate visibility, narratives of *myō* can become perceived as fictitious, if not frivolous. The world of *myō* encompasses all relational dynamics—affiliations, affections, mutual attrition, misapprehension and so forth—that lie outside visible, rational human interactions. Here, conversations with the dead, ghosts, monsters, *yōkai* (supernatural creatures of Japanese folklore), spirits, the unborn and deities unfold. In this context, the pine tree in *nō* theatre stands as one of many symbols in Japan that invites us to practice collective imagination. This dramaturgical object evolves a primordial theatrical exercise, encouraging reflection on what can be seen beyond the visible. The pine tree that is painted at the rear of the nō stage is a mirror reflection of an invisible *yorishiro* (an object via which the gods descend to Earth) that supposedly stands in the middle of the auditorium. The pine tree wall, which

FIGURE 13.2 Okada Toshiki and Kaneuji Teppei's *Eraser Forest*, 2020. Photograph by Keizo Kioku. Image courtesy of 21st Century Museum of Contemporary Art, Kanazawa.

is correctly named the *kagami ita* (literally, mirror board), constantly alerts the audience that their participation in the collective imagination is what maintains (after Merleau-Ponty), the 'invisible' (already including the visible) reality of the pine tree and, thus, the nō performance (Merleau-Ponty 1968, 257). And I surmise that revisiting the Buddhist concept of *myō* reminded by the nō pine tree, is a significantly beneficial practice in philosophically grounding Okada's dramaturgy of 'half-transparent' that emerged from the crucible of Japanese catastrophe-ridden culture (Figure 13.2).

The term 'half-transparent' was conceived by Okada and his collaborator Kaneuji to guide the performers, and sometimes also the audience, in *Eraser Forest* toward a mode of presence that diverges from conventional liveness in theatres. In a world saturated with death and overwhelmed by a diluted sense of presence, exacerbated by the pervasive influence of digital ontologies, they began to perceive the physical presence of performers was simply too pronounced: the unmediated human affect became too in-your-face. To address and amend the representational inequality of too-much-human performers compared to the weaker liveness of various more-than-humans such as plants, animals, ghosts and objects, the artists requested the performers to attain a state of half-transparency. This concept of half-transparency can be illuminated through Hermann Schmitz's notion of 'half-things (*halbdingen*)' (2009, 84). According to Griffero who explains the term by referring to Schmitz's oeuvre, *halbdingen*, comprises elements like silence, sound, music, climate, tension, affect and voice; they are 'neither things nor mere *qualia*' but ontological entities that occupy vast portions of our everyday experience and which 'define the felt body of a person as an epitome of everything that he himself [*sic*] can perceive as belonging to himself, in the vicinity' (Griffero 2017, 38). These intangible, invisible entities infiltrate space and influence our bodies, yet they are seldom recognised as active agents in their own right.

Just as half-things exist at the intersection of visibility/invisibility and tangibility/intangibility, Okada seeks to cultivate half-transparent humans through a more affective ontology. He aims to resist the theatrical hegemony of optics, allowing performers to fade alongside the nonhuman entities that are being demolished or driven to extinction. In this endeavour, he challenges the notion of theatre as a 'vision machine', a critique articulated by Maaike Bleeker (2008, 9). It is essential to note that this dramaturgical decision should not be interpreted literally, not equated with the futile attempt to become nonhuman; for Okada, this is a ludicrous demand that actually confines the ontology of others via human measurement of optics. Instead, Okada and Kaneuji invite performers to occupy a state that resists straightforward calibration—encouraging them to weaken, blur, wither, yield and toggle between; to be liberated from the shells of intransigent individualism and atmospherically relate, resonate and fade together with a multitude of others.

The decision to analyse *Eraser Forest* through the lens of *myō* and half-transparency arises from my first-hand experience attending the performance in Kanazawa. As I observed performers and spectators loiter through expansive exhibition spaces filled with ready-made plastic and assorted junk, it felt as though I was participating in a post-anthropocentric mortuary rite, mourning the imminent loss of humans, nature and, most significantly, divergent objects in the aftermath of neoliberal capitalism. In the white cube setting, both performers and visitors seemed to fade into the background, allowing, in turn, the ghostly entities to express their hidden voices. This central dramaturgy subverts traditional positionality, foregrounding the presence of *myō* beings while making visible humans less present. While engaging in one of the 'EIZO-Theater' pieces—an interdisciplinary theatre format developed by Okada and video artist Yamada Shinpei—I felt immersed in the realm of *jibakurei*, or earth-bound spirits that may not be visible in the physical reality yet can be represented through the video projection. Through multiple screens projecting life-sized images, Okada blurs the distinctions between the alive and the dead, the present and the absent. Reflecting on these experiences, he humorously concludes that he can now confidently assert he has 'absolutely no confidence' in his ability to distinguish reality from illusion in the world (Okada 2022).

Upon entering the exhibition room of EIZO-Theater 1, discerning reality from illusion proves challenging due to the deliberately darkened space that restricts visibility. Gradually, as your eyes adjust to the nocturnal ambiance, you begin to discern an actual-size image of an actor projected onto the rear wall, partially obscured behind a central sculpture. This digital performer takes on the semblance of a *jibakurei* ghost, as the haunting presence is amplified by the diminished status of a dimly lit, digitalised human, in contrast to the monumental installation dominating the room. At the centre rests a gigantic sculpture made of three stacked rocks, resembling a *hokora*, a rustic Shinto shrine often found along rural Japanese pathways where deities and ancestors are believed to reside. This massive *hokora* is infused with life, sharply contrasting with the bloodless performer who fades into the wall to attain a spectral quality. The interpretation of the rock sculpture as a *hokora* is further reinforced by the windmill-like light projected above it, which mimics the spinning motion of the washing machine that the digital male performer claims to have owned for a decade before it went kaput. As the male performer opens his arms and moves them in a circular gesture, he seems to engage in a dialogue with the spinning light, expressing his emotional attachment to the now-defunct appliance. Thus, the sculpture and the light represent not the spirit of a deceased human but rather that of a cherished object—the decade-old washing machine. And the human performer can be described as a *jibakurei* attached to the object, who appears on site whenever an audience enters the room to listen to the repetitive and cyclic 'his-story' of the consumerist world.

During my visit to the museum, I encountered several viewers who peered into the exhibition room and declared, 'the room is quite empty', dismissing the rocks and lights. Others criticised the dim projection, assuming the artist had failed to adjust the installation's lighting properly. Indeed, initially, the rock sculpture may appear insignificant, and the weak video image could seem indicative of artistic negligence. However, when transcending the capitalist logic that instantly overlooks more-than-human agencies, you begin to understand the interplay between video, object, light and the half-transparent performer. This constellation powerfully underscores the relational dynamics of the inanimate object, the spectral human and the haunting atmosphere. By grasping the dramaturgical intention behind the performer's half-transparency, one recognises that the digital performer attains an affective presence, despite—or perhaps because of—the minimal visibility. To reiterate, for audiences to truly perceive existence within the realm of *myō*, they must engage in a non-consumerist mode of spectatorship that requests them to slowly fade into the environment and embrace the half-transparent ghostly agencies.

In conclusion, I have illustrated through Okada and Kaneuji's exhibition/performance that haunted ontologies are vividly experienced through their postvisual dramaturgy. The half-transparent approach invites audiences to collectively engage with what is absent yet palpably present, transcending the confines of everyday visual economy. Dialogues and alliances with entities in the realm of *myō* are affectively foregrounded by dismantling the prevailing human-centric perspective often adopted by museumgoers, which requires unlearning the traditional dichotomy between valued humans and supportive nonhumans. As a mode of agentic resistance to the capitalist culture that habitually overlooks the unrecognised and the unseen, and as a generative response to a society grappling with invisible, protracted catastrophes, the artists render mortal performers half-transparent, reminding and revealing what is fading from our awareness. Through the act of performing fading, Okada and Kaneuji urge us to challenge entrenched human-centric epistemologies, relinquish the status quo and expand our vision beyond the visible—especially as the dead have become, in our contemporary context, the not-so-silent majority.

Note

1 In general, Japanese names are written in the order of the Japanese convention (last name, first name). The names are written otherwise when their works are mainly introduced in Western academia (including myself).

References

Asami, Kazuhiko. 2012. *Genbun & Gendaigoyaku Hōjoki* (*Original Text and Modernized Translation of Hōjoki*). Tokyo: Kasamabunko.

Bleeker, Maaike. 2008. *Visuality in Theatre: The Locus of Looking*. New York: Palgrave Macmillan.

Chen, Mel Y. 2012. *Animacies: Biopolitics, Racial Mattering, and Queer Affect*. Durham: Duke University Press.

Cheng, Meiling. 2002. *In Other Los Angeleses: Multicentric Performance Art*. Berkeley: University of California Press.

Cvecić, Bojana and Ana Vujanović. 2023. *Towards a Transindividual Self: A Study in Social Dramaturgy*. Oslo: Oslo National Academy of the Arts.

Demos, T.J. 2013. *Return to the Postcolony: Specters of Colonialism in Contemporary Art*. Berlin: Sternberg Press.

Eckersall, Peter. 2024. *Dramaturgy to Make Visible: The Legacies of New Dramaturgy for Politics and Performance in Our Times*. London: Routledge.

Glissant, Édouard. 2010. *Poetics of Relation*. Translated by B. Wing. Ann Arbor: The University of Michigan Press.

Gordon, Avery. 2008. *Ghostly Matters: Haunting and the Sociological Imagination*. Minneapolis: University of Minnesota Press.

Griffero, Tonino. 2017. *Quasi-Things: The Paradigm of Atmospheres*. Translated by Sarah De Sanctis. Albany: State University of New York Press.

Halberstam, Jack. 2011. *The Queer Art of Failure*. Durham: Duke University Press.

Harootunian, Harry. 2015. 'Reflections from Fukushima: History, Memory, and the Crises of Contemporaneity', *Boundary 2*, 42(3), 23–35.

Horn, Eva. 2018. *The Future as Catastrophe: Imagining Disaster in the Modern Age*. New York: Columbia University Press.
Iwaki, Kyoko. 2021. 'An Interview with Okada Toshiki'. In P. Eckersall et al., eds, *Okada Toshiki & Japanese Theatre*. Aberystwyth: Performance Research Books, 155–64.
Kobayashi, Kazuhiko. 2013. *Kamo no Chōmei's Hōjoki* (*Hōjoki by Kamo no Chōmei*). Tokyo: NHK Shuppan.
Lavery, Carl. 2016. 'Introduction: Performance and Ecology – What Can Theatre Do?', *Green Letters: Studies in Ecocriticism* 20 (3), 229–36.
Merleau-Ponty, Maurice. 1968. *The Visible and the Invisible*. Edited by C. Lefort. Translated by A. Lingis. Evanston, IL: Northwestern University Press.
Nancy, Jean-Luc. 2000. *Being Singular Plural*. Translated by R.D. Richardson and A.E. O'Byrne. Stanford: Stanford University Press.
Okada, Toshiki. 2016. 'Yurei no sei ha daiji na mondai da (The Lives of Ghosts are Important Issues)', *Viewpoint: The Saison Foundation Newsletter* (77), December, 5–8.
Okada, Toshiki. 2012. *Current location (Genzaichi)* [Performance].
Okada, Toshiki. 2013. *Ground and floor (Jimen to yuka)* [Performance].
Okada, Toshiki. 2016. *Time's journey through a room (Heya ni nagareru jikan no tabi)* [Performance].
Okada, Toshiki and Kaneuji, 2020. Teppei, *Eraser Forest (Keshigomu mori*), [Installation and Performance].
Okada, Toshiki. 2022. 'EIZO-Theater *NEW ILLUSION*', chelfitsch.net. Translate by Aya Ogawa. https://chelfitsch.net/en/activity/2022/07/ni-tokyo.html (Accessed 12 September 2025).
Puar, Jasbir. 2007. *Terrorist Assemblages: Homonationalism in Queer Times*. Durham, NC: Duke University Press.
Rowe, Mark Michael. 2011. *Bones of the Dead: Temples, Burial, and Transformation of Contemporary Japanese Buddhism*. Chicago: University of Chicago Press.
Schmitz, Hermann. 2009. *Der Leib, der Raum und die Gefühle*. Bielefeld: Aisthesis Verlag.
Scholz-Cionca, Stanca. 2021. 'The "Hyper-Real" and the Shadow of NOH'. In Peter Eckersall, Barbara Geilhorn, Andreas Regelbserger, and Cody Poulton, eds, *Okada Toshiki & Japanese Theatre*. Aberystwyth: Performance Research Books, 19–32.
Sedgwick, Eve Kosofsky. 2003. *Touching Feeling: Affect, Pedagogy, Performativity*. Durham: Duke University Press.
Stengers, Isabelle. 2023. *Making Sense in Commons: A Reading of Whitehead in Times of Collapse*. Translated by T. Lamarre. Minneapolis: University of Minnesota Press.
Sueki, Fumihiko. 2019. *Myōken no tetsugaku 2: Ima nihon kara okosu tetsugaku* (*The Philosophy of Myōken: Philosophizing from Japan Now*). Tokyo: Pneuma Sha.
Uchino, Tadashi. 2020. 'What Is: "Theatre That Does Not Speak to the Audience"?—On Eraser Mountain.' In Kurosawa Hiromi, Nakata Koichi, Ikeda Ayumi, Itabashi, Mari and Ogami Takashi, eds, *chelfitsch & Teppei Kaneuji Eraser Stone*. Kanazawa: Shukyu and 21st Century Museum of Contemporary Art, 76–85.
Yoshida, Kenkō and Kamo no Chōmei. 2013. *Essays in Idleness and Hōjoki*. Translated by M. McKinney. London: Penguin Classics.

14

PERFORMING ACROSS TIME

Jill Orr in Conversation with Peta Tait

Jill Orr and Peta Tait

Note: This chapter introduces readers to Jill Orr's extensive environmental performance which has been created consistently over decades. She continues to provide influential precedents in durational live performance and photography. In particular, Orr's body based works—which are performed naked or costumed—led to her reputation as an international leader in performance about human connections with ecological systems. Orr continues to perform in outdoor ecologies and to encourage younger ecologically focussed artists.

Jill acknowledges the Australian Aboriginal Indigenous Kulin Nation, the Wurundjeri country that we meet on.

Peta: Your early ecological performance series, *Bleeding Trees* brought your work to national prominence in Australia in 1979 and it has become particularly well known for complex meanings that associate human suffering with that of trees which are cut down. Anne Marsh writes that your performance 'drew attention to the devastation of trees' but replaced a 'dead tree' with the 'living body' of the artist (1993, 121). Can you explain your approach to creating this work?

Jill: My approach for this work was to use the body as an emotional barometer for trees, in empathy with trees. The body is a stand-in for trees. But, of course, in the 1970s, it also had the lens of the early feminist critique. And so in this sense, the body was seen as pandering to the male gaze but when we put an environmental lens on it, the environment lens is upon all regardless of gender and species.

Peta: Marsh continues that the positioning of the body in this 1979 series gives the impression of 'an open wound' and this is emphasised by

DOI: 10.4324/9781003584445-19

the image of your buried head and open mouth. She also describes how you present ideas of silencing speech in your work.

Jill: The power of an image, performed live or photographically, is not always bound by language, sometimes it bypasses and communicates viscerally and bodily. In this way there is a 'silencing of speech' which enables or triggers a gut response. That is an effect of art.

Peta: I find the way in which your early work has been interpreted as like ritual and potentially shamanistic in resonance interesting but, perhaps, less important than the overriding environmental significance. In 1979 *Lunch with the Birds* is set on a beach with seagulls and small fish. You routinely locate your performance in outdoor environments and this has become a distinguishing feature of your work over decades.

Jill: We are embodied creatures bound to place and directly or indirectly dependent on natural environments both physically, psychologically and spiritually. I choose sites that have an historic and environmental significance and are awe inspiring. In that way the performance is in collaboration with the site where the connection between body in place can 'speak' through the language of art.

I counter how we are in our minds—although that's a great place to be because our mind registers our heart. The body being fully present is central to performance and training methods like Butoh and Body Weather movement practices (Marshall 2024). Improvisation, walking and drawing have evolved since the 1960s and 1970s and are still practiced, adapted and renewed over time (Figure 14.1).

Peta: Some of your works are clearly focussed on gender identity in what might be termed durational ordeal performance. For example, the performance in which your hair is attached by chains to the gallery ceiling and ceremonially cut by the audience in *She Had Long Golden Hair* in 1980. Again with your first gender ambiguous figure in white in *Love Songs* in 1989. Other works suggest a preoccupation with time and the passing of time. For example, with your body suspensions in the air and, in particular, the one involving your body's suspended from ropes and balanced on ice with a fire underneath to melt the ice for *Pain Melts 1* in 1980.

Jill: Time's impact is subtle, molecular, geological and cosmic. Time impacts our bodies, human and non-human alike. It impacts our lives constantly. Time is an element of performance that is like paint for a painter, it is the medium through which actions occur and over time change becomes visible.

FIGURE 14.1 Jill Orr, *Walking on Planet Earth*, 1989. Photographer: Virginia Fraser. Image courtesy of Jill Orr.

Peta: As Marsh explains, your early 1970s works might be interpreted within the art tradition of landscape even though the 'political-ecological tendency' was unmistakeable by 1989 when you make *Walking on Planet Earth* with your figure costumed in a full length dress as a colonial woman confronting a bulldozer (1993, 128). She is photographed with a burning umbrella.

Jill: The late Virginia Fraser photographed that work which was performed in Clifton Hill [Melbourne]. This place is now a fantastic park, and reserve. In 1989, however, it was a rubbish dump that was being transformed with earthworks that resembled outback Australia. Imagine it's hot, it is so hot that your very protection [umbrella] bursts into flames. *Walking on Planet Earth* is also my early imaginings of the experience of colonial women. And of course, beside it—but I could not go directly there—was the Indigenous land that we have invaded, damaged, and those sort of issues are intertwined.

Peta: You return to the potent image of a burning umbrella in a later work because the image has clearly become even more prescient with climate change.

Jill: The *Southern Cross – To Bear and Behold* in 2009 revives the image of what could be read as a colonial woman and a burning umbrella and adds to it with other colonialising figures. It's 30 years after my first *Walking on Planet Earth* burning umbrella. And I thought, I will do the same action on Mitre Lake. Mitre Lake sits at the base of Mount Arapiles in the Wimmera [state of Victoria in Australia]. It's a natural salt lake. It has a surface that's crustaceous salt, but when you walk on it, an oozing sulphuric mud is revealed leaving an imprint. Creatures go across this lake that's not a lake. They are deceived in the centre, and you can see the tracks as they struggle to either get out of the centre or die.

The traditional custodians are the Djurid Baluk clan of the Wotjobaluk people (Figure 14.2). When I stood on Mount Arapiles and looked down at Mitre Lake below, I could see vast country for miles. It is now wheat farming country. The tractors plough, turn around the lone tree and up round the fence-line. It is an incredible pattern of farming. But the thing is, in standing there, up on the top of the mountain, one can actually imagine what is underneath the farming. One can feel this ancient place. One can respect the Indigenous understanding.

FIGURE 14.2 Jill Orr, *Southern Cross To Bear and Behold: Burning*, 2007. Photograph by Naomi Herzog. Image courtesy of Jill Orr.

Peta: The Mitre Lake performances about damage done to the land also include the image of the priest-like figure dressed in black walking across the salt lake carrying a large full sized cross on their/his back, and leaving the grey dents of footsteps on the surface. This is one of my favourite images from your work and I bought a small photograph. As Marsh says, you have produced 'extraordinary performance art documents in the form of photographs' that expand on your live performance (2014, 67).

Jill: This lake is like a mirror, reflecting both the ancient and recent past and potential future. It's full of implicit presence and absence. Fire, in all its destruction, is also symbolic of hope. I was thinking of hope as the spirit for the future. Hope, as a flicker of imagination. Hope as a way to keep motivated, offering signs of possibility. And hope as an act of resistance. But in a recent conversation, with you the other day, Peta, we talked about hope, and you said, 'I think it's trust in living in the future'. But I think you actually need that substructure of hope as the inspirational underpinning in order to hold trust.

Peta: My other favourite image from your work is a large photograph from *The Promised Land.* You are costumed in light coloured clothing that makes your gender and your ethnic and historical identity ambiguous as you stand in the skeleton of a boat, paddling on the water's edge like a boat person or refugee (Tait 2018). It has a timeless quality. I saw you perform in Venice in 2012 in the Palazo Bembo in a large room on the main canal packed with hundreds of spectators, who were completely silent and focussed on your action, and included other leading international performance makers.

Jill: I think, as you develop as an artist, you gain a little confidence, or you gain a modicum of self-awareness, but not enough. To make a work there must be something deeply personal and simultaneously something that connects beyond the self to others. It is a communal thing. The vital purpose is the interconnectivity. Essentially it is trust that there is an energetic connection that goes beyond this room, beyond the horizon. I need a deep personal motivation as well as, equally, a broader collective relevance to develop a work.

Peta: I have explored your capacity to evoke a strong bodily response within a phenomenological framework and particularly your confronting works with dead animals about human-animal relations (Tait 2015). But it is your photographic imagery of the human figure in an outdoors setting that is most environmentally meaningful and the image often encompasses a physical atmosphere that creates an emotional mood.

Jill: I am going to clouds enveloping the globe to answer this in a left-handed sort of way. Clouds don't stop, they change in density, volume and interact in constant spatial relations. This molecular exchange doesn't stop because it's in Australia, South Africa or Europe etc., it is an all enveloping atmospheric action that surrounds the Earth. That's phenomenal because it is our finite water resource keeping Earth alive as we know it. We feel extreme heat as the skin signals melting and the core temperature of everything is unbearable. Weather extremes become enforced embodiment experienced by all in the vicinity. For me, performance hopes to communicate through a shared corporeality.

Peta: I find myself returning to what I think of as your classic series, *Antipodean Epic* from 2015 to 2016 because it is so full of interpretative possibilities (Tait et al. 2025). It was performed for Palimpsest, Mildura's biannual art exhibition curated by Helen Vivian, in which she encouraged a local environmental focus. Local but contradictorily national because the region encompasses the Murray-Darling river system that flows down Australia's east coast through four states.

Jill: *Antipodean Epic* incorporates photographic, video and live performances created on three sites round Mildura. The first video/photographic performance was on a massive wheat farm, and focussed on a bird like character, The Interloper. The second site was in the Sunset country, where the straw creature emerges from acres of gypsum mining trailings that are dumped there. The final site is the disused railway building in Mildura, where the videos give context and signal the arrival of the creatures in the live performance. The creatures are ambiguous because each related to completely different environments. The wheat farm, a productive landscape using massive harvesters. I asked the farmer: 'Would you mind very much if you got your great big harvester with 10 tonnes of wheat and poured it all over me, please?' And he said, 'Oh, that'll only take 30 seconds'. And I said, 'Do you think you could do it in loads of 10? A little bit slower. 10 seconds, 10 seconds, 10 seconds'. And he said, 'Okay'. So, anyway, that is the image. I just gently put a bit of skin in, oh, that doesn't hurt, that's okay. So I just let it go, covering my mouth and nose in the process. The Interloper is buried in wheat and slowly emerges.

Peta: So the weight of the wheat pouring into a storage container from the harvester with your bird figure in between—your skin—was choreographed. The bird is strange almost uncanny but at the same time, engaging and quirky. Can you talk about what inspired this figure.

Jill: The question is, are the creatures at the end of their species or are they at the beginning of another? Or are they displaced or transported viral creations? This is before COVID. Are they unwanted interlopers within the seed stock? Are they the carriers of a potential future or remnants of a distant past? Or both? I named it—no, I named them, the Interloper, who really looks like a serious misplaced person in drag. They frolic. They become hunted. They are taken away swinging on a pole like a hunted animal and finally reappearing suspended in the railway shed witnessed by the dust-suited audience. I won't go into the remainder of the work here, only to say that rather than a stand-in for creatures, the creatures become animal-human hybrids.

Peta: In 2021 you return to your earliest environmental inspiration in *This Tree* in a photograph series taken by Christina Simons. You performed near a highly trafficked freeway where there is a large tree stump that was cut off above your height.

Jill: *This Tree* is really becoming a tree, a bit like a human-tree hybrid. So I think, in a way, I've come full circle from 1979 to 2021/22, only instead of standing in for the tree, the endeavour is to merge with the tree. *Jill reads from a statement about this art work.* This tree was magnificent. This tree, a testament to progress, whose wake is unforgiving…. This tree is a body that is changing. This tree, grafting between species. This tree as home. This tree stands on Wurundjeri country.

Peta: Thank you so much Jill for your performance that has inspired us to think and feel ecologically over decades.

References

Marsh, Anne.1993. *Body and Self: Performance Art in Australia 1969–92*. Melbourne: Oxford University Press Australia.

Marsh, Anne. 2014. *Performance Ritual Document.* Melbourne: Macmillan.

Marshall, Jonathan. 2024. *Butoh and Suzuki Performance in Australia.* Leiden: Brill.

Tait, Peta. 2015. 'Fleshing Dead Animals: Sensory Body Phenomenology in Performance.' In Maaike Bleeker, Jon Foley Sherman and Eirini Nedelkopoulou, eds, *Performance and Phenomenology.* London: Routledge, 111–20.

Tait, Peta. 2018. 'Performing Ghosts, Emotion, and Sensory Environments.' In Lara Stevens, Peta Tait and Denise Varney, eds, *Feminist Ecologies: Changing Environments in the Anthropocene.* Basingstoke: Palgrave Macmillan, 175–91.

Tait, Peta, Denise Varney, and Lara Stevens. 2025. *Ecology and Climate in Theatre and Australian Performance.* Cambridge: Cambridge University Press.

15

DRAMATURGY FOR COMMUNITY RENEWAL AND ENVIRONMENTAL CARE

Linda Hassall

The dramaturgical framework discussed intends to assist artist-researchers and theatremakers working in disaster-affected communities. The term artist-researcher refers to practitioners working in universities and whose main research activity is artistic practice.[1] The term theatremaker refers to artists who usually make theatre outside academia such as actors, stage and production managers and sound and lighting designers. The research design incorporates collaborations between both groups, and with the affected community. Working together, they offer an environment where shared experiences become a cornerstone of recovery. What is learned from the collaboration becomes the research.

The model offers a scaffolded process framed by ecological thought and design. Using principles of ecodramaturgy, climate dramaturgy and ecoscenography (ecosceno), this framework considers positive environmental communication and action in disaster areas in site-specific contexts.[2] Referred to as ecosceno dramaturgies (Hassall et al. 2023)—it can be applied when working in tandem with local communities and organisations on collaborative creative strategies for climate-change communication, activism, adaptation and equity (Brown et al. 2017). The framework is particularly useful for developing site-specific performance in disaster-affected communities, as 'place' itself, 'is the dramatic provocation, a space of resistance, response, recovery, and renewal' (Hassall et al. 2023). Site can disrupt the usual spatial transaction between performance and audience (Pearson 2017, 296). Making theatre on the site of a climate disaster provides agency for artist-researchers, theatremakers and the community alike to undertake new collaborative transactions and to experience transformative encounters within the site itself.

DOI: 10.4324/9781003584445-20

This occurs between the actual place (site) and its regeneration, and the psychological space, which contains the climate experience and its after-effects. As a systemic dramaturgical approach, ecosceno dramaturgies consider the actual, physical place and the psychological space as part of the social and the cultural environment of the community living with climate-change experiences. Considering all stages of the theatremaking process from initial site visits to post-production extraction, the framework proposes a unified eco-creative process for facilitating aspects of community connection and renewal. As a malleable practice, ecosceno dramaturgy responds to calls for developing 'creative artistic theatre practice in the context of living through the rapidly accelerating crisis of global climate breakdown' (Svendsen 2023, 6) and for climate-change responses that 'embrace the complexity, mutability, interconnectivity, inconstancy and the uncertainty of ecology' (Benner 2022, 279).

The research sits at the interface between social science and applied theatre research (ATR) and reflects on how collaborators can balance aesthetics and ethics in partnership with vulnerable community groups of people who are being asked to share and/or stage their experiences of disaster (Sadeghi-Yekta et al. 2023). The framework acknowledges that artists who create theatre within vulnerable communities consider the difficulties of developing the structures in which communities are invited to share their experiences and to fully contemplate the ethics of doing so as a paramount concern in the theatremaking process (Thompson 2015, 433–36).

Originally conceived for the Flood Project (2023–24), the dramaturgies were developed to situate real life testimonies of a flood-affected community within a climate disaster site in the suburb of Rocklea in Meanjin, Brisbane, Australia.[3] The suburb was described in the media as Ground Zero, giving an indication of the devastation caused to homes, businesses and community facilities by widespread flooding in 2022. The ecosceno dramaturgy responded to the devastation and emerged through a research process derived from the application of ATR methods. In attempting to re-narrate the disaster site, it aimed to share community experiences about recovery while addressing how performance making could promote environmental recovery. In doing so, the site resonated multiple layers of experiences including trauma and renewal and social activism and awareness.

Supporting principles of slow dramaturgy as discussed by Peter Eckersall and Eddie Patterson (2011) and Carol Martin (2010) the dramaturgies are applied in 'slow time' with community participants. Encouraging the sharing of life experiences, inspiring authentic recovery stories of resilience, activism and agency, the significance of fostering collaborative relationships over time based around trust and respect is paramount (Eckersall et al. 2011, 179). In highlighting the human stories behind the disaster, the dramaturgies can humanise and recontextualise these experiences. It suggests that 'community resilience is at the heart of the developing work and any production ultimately respects the communities *ongoing* experiences of trauma, including psychological well-being and environmental recovery' (Hassall et al. 2023). The framework outlines a relationship between human-centred agency and more-than human ecological sensibilities. Human-centric issues can be told from experiential perspectives and can therefore connect with the climate-change experiences of the community. Sometimes giving someone an opportunity to share their experience can make a huge difference in the recovery cycle.

Ecosceno Dramaturgical Framework

The United Nations has identified climate change as the 'defining crisis of our time' and suggest that 'as the infinite cost of climate change reaches irreversible highs, now is the time for bold collective action' (United Nations 2019). Offering an explicit suite of philosophical, ideation and practice considerations for making theatre in climate-affected sites, ecosceno dramaturgy is a collaborative dramaturgy that aims to enrich the experiences of communities exposed to eco-trauma. The framework ensures dramaturgical sensibilities cohere with sustainable processes and considers the eco-creative potential in sites of eco-trauma. ATR methods dictate that community narratives are offered freely and without coercion. Sharing real experiences of the disaster and post-disaster recoveries encourage authentic dialogues that potentially form the basis of climate-theatre performance. While this framework was developed for a specific flood-affected community in Meanjin, Brisbane—after the flood event in 2022, the framework is malleable, acknowledging every climate event is unique and every community experiences disaster from their own perspectives. The framework can be used as is or in an explicit way to compliment dramaturgical structures in place in site-specific contexts. Ultimately the collaborators can apply the framework as required by community needs while considering the relationship between ecological well-being and eco-practice applications. Ecosceno dramaturgy includes the following phases.

Eco-Analysis: Past and Present Climate Change Encounters in Space and Place

The influence of climate science on the Flood Project is not restrictive. Using climate-change evidence to inform and compliment the practice can be useful. Climate change and climate disaster propose that the world, life and human relationships with nature are essentially fluctuating, complex and improvised (Szerszynski 2003; Szerszynski et al. 2003), and this can be acknowledged in the theatremaking. Analysing the impact of past catastrophes, the continuing disruption to the site as well as ongoing environmental occurrences is one of the first steps in devising a project with affected communities. The Rocklea site, the subject of the first application of this framework, was also the site of a more devastating flood event in 2011. Many community members were still recovering from this event. It is vitally important that during this early research phase, everyone involved in project development is sensitive to past eco-traumatic experiences and the possibility of ongoing climate-anxiety in the community. Questions around trauma, memory and healing arise through individual experiences and inform the overall process. Such questions include:

- How can a theatre project applied in a site of ongoing eco-trauma investigate truth of experience to support community transition, change and healing?
- How can drama/theatre be used as an effective pedagogy to address the dramatic re-narration and theatrical re-construction of site (place) as a site of recovery?

The framework can negotiate the tension between drawing on verbatim material and reimagining experience. The alliance between artist-researchers, theatremakers and community participants also aims to capture the joyful, beautiful and hopeful encounters that have arisen.

Eco-Connectivity: Foster Collaboration with People in Place for Community Engagement

Community engagement is not an end to eco-trauma, but it can be fundamental to complete recovery (EMC 2014). Eco-creative engagement can assist communities to make decisions about their ongoing recovery or can help individuals analyse eco-trauma as part of a communal rather than individual experience. Establishing authentic connections within the community either as direct or second-hand experience (emergency services and disaster support), is essential to this phase of ecosceno practice. The early connections made support discovery and learning while offering the community agency through collaborative engagement processes. Significantly early engagement enables artist-researchers to explore issues of concern to communities, to identify problems which will actively inform building recovery narratives or offer possibilities to provoke social change and or promote education and awareness. Creating an engagement plan to inform the process is useful prior to entering the community and can include:

- Develop constructive dialogue plans for communicating with the community.
- Advise the community on project planning, anticipated delivery and possible levels of participation.
- Involve the community on all matters that affect them pertaining to the production.

- Collaborate with community at all levels of project to empower them to engage as project experts.
- Facilitate timely and respectful engagement processes that inform eco-creative development.
- Value diversity and different levels of eco-traumatic experiences.
- Schedule community engagement tasks to provide maximum participation opportunities for participants to collaborate on and in the project.
- Document everything. Keep a cyclic feedback loop in place with participants throughout.[4]

Eco-Alliance: Establishing Trust for Requesting Stories of Eco-Recovery

Establishing trust is achieved by respecting multiple layers of impact within community experiences. It is important to ensure that community members have adequate time to consider and prepare their responses to the project and this may mean sharing questions with them prior to undertaking any interpersonal dialogues—working in slow dramaturgical processes that place the participant front and centre of the ecosceno dramaturgical alliance. Artist-researchers and theatremakers must always remain accountable by providing community members with a clear understanding of how their valuable insights, experiences and input will be used. It is ethical practice to ensure a feedback loop is practiced throughout so responses are continually assessed to foster and embed productive alliances between all concerned. Ensuring any communication with a climate-affected community—for example, initial discussions, the conducting of interviews for collating climate-change narratives and the development of verbatim text for performance—is ethical.

Eco-Emotional Well-Being: Acknowledge the Site as Ongoing Psychological Space

The impact of being in a climate disaster space that contains historical eco-trauma can foster psychological impacts due to experiencing that place long term. Black Dog Institute (2024) posits that up to 50 per cent of persons living through unpredictable and extreme weather events can develop immediate elevated risks of mental health, and up to 20 per cent of these persons may experience PTS in the years following the event.[5] The complexity of these issues suggests dialogues for encouraging co-participation in the project are built transparently and respectfully around the value of each contribution, large or small. Risking intense emotional responses from participants should be avoided at all costs. Climate psychologists from the Climate Psychology Alliance (2020) identify the escalation of heightened emotional, mental or somatic distress in response to dangerous changes in the climate system.[6] They suggest that healthy responses include talking and listening about experiences, fears and concerns to encourage recovery dialogues. Never ask the community participants to re-live trauma associated with the disaster but rather focusses on their agentic experiences and circumstances of recovery and resilience.

Eco-Narratology: Presentation of Narratives

Verbatim performance is done in collaboration with the community. Artist-researchers and theatremakers are cast amidst untold stories, and it is our responsibility to provide space and air for these stories to be shared. Thoughtful and respectful consideration and continued communication about how narratives can be shared as performance in site is essential. ATR acknowledges the site as a place for incorporating the multiplicity of alternative ways of knowing. In doing so, the site transforms to a place of eco-narration. Through eco-narratology the space becomes a place of flux, wherein stories presented respond to the materiality, the history and the disturbance of experience. Drawing on Kown's (2002) site-specific art making theories, the eco-narratology becomes a form of performance that is site-determined, site-oriented, site-referenced, site-consciousness and site-related.

Eco-Sensitivity: Treading Softly in Human and No-Human Encounters

It is likely that community members will invest and engage in the creative development processes and even participate in the performance, as ATR projects are aimed at speaking with the community rather than for the community. Ensuring artist-researchers and production teams are sensitive to the human participants' experiences of and in the site is a significant part of the ethical considerations framing the theatremaking. Further ensuring that the ecosceno dramaturgies make the least impact on the ecology of the site is equally important. Essentially, the performative development is constantly reshaped and adapted according to the daily observations of the sensitive needs and interests of the community and *of the site itself.*

Ecoscenography: Ecological Production vs Over-Consumption

The framework encourages ecological theatre that champions sustainability as the heart of production process and is the eco-creative extension of the ecosceno dramaturgical process. Engaging with sustainable production design, significantly enhance the storytelling experience and the nature of the audience experience. From an eco-ethical position, this stage supports appropriate design processes informing the overall aesthetics of the work. Intrinsic to the process is Tanja Beer's ecosceno, a design practice that challenges traditional approaches to performance design (2021, 101–38). Applying Beer's ecoscenographic approach guarantees design practices consider protecting nature as ultimate to the process. As a holistic approach, it respects multi-scalar engagement with the site through what Beer calls a 3C approach: *co-creation*—a place-based community approach to all design concepts, *celebration* of eco and sustainable approaches and *circulation*—sharing philosophies, ideas, approaches and redistributing resources and materials at end of project to support community recovery.

Eco-Creative Empowerment

Eco-creativity acknowledges the power of sharing emotional (subjective), embodied (practice) and artistic (creative) understandings of socio-cultural and historical experiences of climate change in the specific site. Eco-creative approaches are a necessity for empowering communities for dealing with environmental problems in innovative ways and foster pathways to resilience and recovery. Developing eco-creative responses to climate-change experiences requires researchers and theatremakers to be ecologically empathetic in all aspects of the project.

Eco-Encounters: Engagement with Space, Place and People

As a site-specific performance project, audience experience can be identified as an eco-encounter. Climate-change disaster can sometimes foster an 'us and them mentality' for example those who have experienced disaster firsthand and those who have only experienced it second-hand via news programs, weather reports or social-media posts. In some places, it may be that the event only affected specific areas of a city or community, while some areas were left completely unscathed. This occurred in the 2022 Meanjin, Brisbane flood, wherein most of the city, specifically the North side of the Brisbane River, was completely unaffected. Northsiders mostly had little if any experience of the devastation affecting their Southside neighbours apart from mediatised bites of local news or similar. The eco-encounter therefore can become a site for broader public discourse. The encounter between a spectator audience who experiences the climate disaster through the theatre experience can significantly reframe their understanding (or lack thereof) of the disaster through the eco-narratology. Their eco-encounter can raise ecological consciousness about climate-change events and/or inspire curiosity about the ecological well-being of place. As such the eco-encounter inspires understanding, meaning, empathy and effect beyond the disaster itself.

Eco-Extraction from Place and Space

By embedding the ecoscenographic framework, within muti-layered perspectives, impact is identified beyond merely human issues. The ecosceno dramaturgy acknowledges not just the importance of developing deeper, sustainable and meaningful relationships with the community, but with the land on which the community lives. Eco-extraction considers the environmental impacts of leaving the site in the same state as when first encountered. Artist teams might perform carbon off-setting activities in the space post-production. This may be done in association with local councils and include several mitigating activities dependant on the needs of the site and the community.

Renewal

As climate change escalates in the contemporary consciousness, the processes of working in sites of eco-trauma enables artist-researchers and theatremakers to encourage climate activism using responsive dramaturgies as a platform to raise awareness, inspire action and envisage more sustainable futures. We can actively inform approaches to resilience planning between communities and local Councils. Community organisations can understand the needs of climate-affected communities and can consider restructuring policies and programs to address the needs of those vulnerable to environmental hazard (Vilá et al. 2023). When working in locations of specific eco-trauma, the ecosceno dramaturgical framework provides us with an opportunity to drill into the experiences from perspectives of resilience, renewal and agency and proposes a transformative encounter between physical place—the site and psychological space—and the experience.

Notes

1 See Zoe Svendsen's (2025) research based directorial, design and dramaturgical work with METIS in the United Kingdom. She is also Lecturer in Faculty of English at Cambridge University.

2 See Lisa Woynarski's work on eco-dramaturgy (2020). Ecodramaturgy is a critical framework for making/studying theatre encompassing environmental narratology and sustainable producing practices. Zoe Svendsen's work on climate dramaturgy (2023). Climate dramaturgy is making climate-positive and carbon-reducing choices in theatre-making. Tanja Beer's work (2021) on ecoscenography. Ecoscenography embeds principles of ecology to create recyclable, biodegradable, restorative and/or regenerative performance spaces.

3 Meanjin is a Yuggera and Turrbal word, referring to the area where central Brisbane now sits. The word can also refer to the broader Indigenous country encompassing Brisbane. See Hassall et al. (2023) for community regeneration.

4 Governments and Councils have community engagement plans to inform post disaster recovery. Artists can use these to assess community priorities and adapt to inform the eco-creative processes.

5 The Black Dog Institute is the only medical research institute in Australia that investigates mental health across the lifespan—from childhood to adulthood.
6 The Climate Psychology Alliance is a North America organisation that addresses the urgent psychological dimensions of the climate and ecological crisis and promotes cultural shifts toward human resilience, regeneration, and equity. They promote education and social inclusion and advocacy and offer mental health services to those affected.

References

Beer, Tanja. 2021. *Ecoscenography: An Introduction to Ecological Design for Performance*. Singapore: Springer Nature.

Benner, Saffron. 2022. 'Driving 'Transformational Change: Using Ecodramaturgy to Develop a More Sustainable Theatre Ecosystem.' *Australasian Drama Studies* (80), 278–303.

Black Dog Institute. 2024. 'The Implications of Climate Change on Mental Health.' https://www.blackdoginstitute.org.au/news/the-implications-of-climate-change-on-mental-health, accessed 14 October 2024.

Brown, Katrina, Eernstman, Natalia, Alexander Huke, and Nick Reding. 2017. 'The Drama of Resilience: Learning, Doing, and Sharing for Sustainability.' *Ecology and Society* 22(2). https://doi.org/10.5751/ES-09145-220208.

Climate Psychology Alliance. 2020. 'The Handbook of Climate Psychology.' www.climatepsychologyalliance.org/index.php/component/content/article/climate-psychology-handbook?catid=15&Itemid=101%20, accessed 13 September 2024.

Eckersall, Peter and Eddie Paterson. 2011. 'Slow Dramaturgy: Renegotiating Politics and Staging the Everyday.' *Australasian Drama Studies* (58), 178–92.

EMC (Emergency Management Victoria). 2014. 'Engaging the Community in Disaster Recovery, Disaster Recovery Toolkit for Local Governments.' www.emv.vic.gov.au/how-we-help/disaster-recovery-toolkit-for-local-government, accessed 11 October 2024.

Hassall, Linda. et al. 2023. 'The Flood Project: Thinking through an Applied Dramaturgy Framework for Community Regeneration.' *Critical Stages/Scènes Critiques* (28). www.critical-stages.org/28/the-flood-project-thinking-through-an-applied-dramaturgy-framework-for-community-regeneration/, accessed 6 September 2024.

Kwon, Miwon. 2002. *One Place after Another: Site-Specific Art and Locational Identity*. Cambridge: MIT Press.

Martin, Carol., ed. 2010. *Dramaturgy of the Real on the World Stage*. London: Palgrave.

Pearson, Mike. 2017. 'Site-Specific Theatre.' In A. Aronson, ed., *Routledge Companion to Scenography*. London: Routledge, 295–301.

Sadeghi-Yekta, Kirsten, Monica, Prendergast, and Michael Balfour, eds. 2023. *Applied Theatre: Ethics*. London: Methuen Drama/Bloomsbury.

Svendsen, Zoe. 2023. *Climate Conversations: Making Theatre in the Context of Climate Crisis*. Report. https://hightide.org.uk/climate-conversations-a-climate-dramaturgy-research-report/ 11 September 2024.

Svendsen, Zoe. 2025. 'About Zoë.' *METIS*. https://metisarts.co.uk/collaboratorszoe-svendsen, accessed 16 April 2025.

Szerszynski, Bronislaw. 2003. 'Technology, Performance and Life Itself: Hannah Arendt and the Fate of Nature.' *Sociological Review* 52 (2), 203–18. https://doi.org/10.1111/j.1467-954X.2004.00459.x.

Szerszynski, Bronislaw, Wallace Heim, and Claire Waterton. 2003. 'Introduction.' *Sociological Review* 52 (2), 1–14. https://doi.org/10.1111/j.1467-954X.2004.00447.x.

Thompson, James. 2015. 'Towards an Aesthetics of Care.' *Research in Drama Education: The Journal of Applied Theatre and Performance* 20 (4), 430–41. https://doi.org/10.1080/13569783.2015.1068109.

United Nations. 2019. 'The Climate Crisis: A Race We Can Win.' www.un.org/en/un75/climate-crisis-race-we-can-win, accessed 11 September 2024.

Vilá, Olivia, Bethany Cutts, Whitney Knollenberg, and Louie Rivers. 2023. 'Environmental Justice in Disaster Recovery: Recognition of the Latinex Community by No-Profit Leaders.' *Climate Risk Management* 40. https://doi.org/10.1016/j.crm.2023.100502.

Woynarski, Lisa, 2020. *Ecodramaturgies: Theatre, performance and climate change.* Springer Nature.

16

PERFORMING CLIMATE ACTIVISM

Deborah Hart with Denise Varney

Intermediaries

The Climate Guardians are an activist ensemble with a captivating but deadly serious performative interpretation of the art of political protest. Their practice is to appear in public dressed as angels in long white robes, large, feathered wings and garlands in their hair for halos. The Angels tend to be women, multi-aged, middle-classed and committed, numbering from 6 to 60 at any one action. Their acts consist of flocking to selected sites to bear silent witness to the perpetrators of climate change including stock exchanges, banks, corporate offices, government buildings and ports. Their mission is to protect ecologically endangered sites: forests, rivers, seascapes, native animals and the atmosphere in the tradition of protest through durational silent witness. The angels aim to stir a guilty conscience among polluters at the same time as they offer an appearance to support protests such as the School Strike for Climate in the City Melbourne (see Figure 16.1).

Origins

The Climate Guardians trace their genesis to a vigil enacted by Tasmanian performance artist Allana Beltran, whose embodied 'Weld Angel' appeared in an old-growth forest in the Weld Valley, Tasmania, in 2006 (see Figure 16.2). Dressed in a long white dress with large, hand-made feathered wings, Beltran harnessed herself between two ancient trees on a tripod ten metres above ground level. Her ten-hour silent protest succeeded in temporarily halting logging in the old-growth forest. Beltran's performance was ethereal to watch. It captured

DOI: 10.4324/9781003584445-21

FIGURE 16.1 Climate Guardians outside the Old Treasury Building in Melbourne during the School Strike 4 Climate protest, December 2018. Image courtesy of Julian Meehan.

FIGURE 16.2 Allana Beltran as the 'Weld Angel', 2007. Photograph by Matthew Newton. Image courtesy of Matthew Newton.

many of the same dimensions as the famous images of protestors at the Franklin River in 1983. Beltran's concept of a singular 'solitary stance' expanded into hosts of angels (Varney 2019, 67).

Public

Deborah Hart recalls one of the early Climate Guardians appearances (Figure 16.3):

> We were dressing up as angels to go out on the streets to fight for climate justice. We staged actions such as stencilling a BHP plinth with graffiti that read 'climate justice'. We stood in line, holding hands as security went nuts. The police came and said, 'Ladies, you look so lovely.' They posed for photos with us before suggesting that we actually stencil the footpath in front of BHP in order not to vandalise private property. BHP security said, 'Oh, my God. Arrest them. Arrest these women'. That was our first experience as the Climate Guardians. On the way home on the tram, the tram driver said, 'I want everybody to stop and acknowledge the angels because we have never, in all my years of driving a tram, ever had a flock of angels hop on.

FIGURE 16.3 Climate Guardians delivering a message to BHP's CEO and Board at its global headquarters in Melbourne, 2019. Photo courtesy of Julian Meehan.

Community Interventions: Gas Hub at Crib Point

Deb Hart recalls:

> The Climate Guardians visited Crib Point on Western Port Bay, south-east of the city of Melbourne on 22 February 2021. The visitation was strategically timed to coincide with the Victorian Planning Minister's review of the findings of a Public Inquiry into Australian energy company AGL's proposed 300 metre floating liquid gas hub. Artists and a marine biologist joined the flock of seven Climate Guardians, who were welcomed by a crowd of around 30 Save Western Port community members. The local *Mornington Peninsula News* sent a journalist and a photographer to cover the Guardians' performance with one of the images (see Figure 16.4), used on one of

FIGURE 16.4 Climate Guardian Janenne Willis at Crib Point, Victoria, protesting against a proposed gas terminal, 21 February 2022. Photograph by Julian Meehan. Image courtesy of Julian Meehan.

three postcards distributed by Environment Victoria. Out of a total of 3,000 postcards presented to the Minister's office with handwritten notes, it was estimated that 1,000 were from people who attended the Climate Guardians' performance.

On 30 March 2021, the Victorian Government rejected AGL's proposal on the grounds that it would have 'unacceptable environmental effects' on a UNESCO listed wetland of international significance (AAP 2021). On 9 April 2021 AGL withdrew its proposal. After four long years and hundreds of millions of wasted dollars, the local community finally defeated an appalling project. The Climate Guardians were thrilled that showing up to offer solidarity at a key time of their campaign at least gave the Save Western Port community a well-deserved lift.

Paris: 2015 Climate Change Conference (COP21)

Hart reflects:

> The Climate Guardians visited Paris for COP21 in 2015, taking with us hundreds of beautiful artwork messages from school children. In Paris, we read

FIGURE 16.5 Climate Guardians perform protest at the United Nations Climate Change Conference COP21 in Paris, 2015. Image courtesy of CLIMARTE.

> these out and displayed them on the streets, again dressed as angels. We had this real mission to take the voices of the people whose lives would be profoundly impacted by the decisions being made at the conference. We were breaking the state of emergency laws following the recent terrorist attacks, but the French police really left us alone. We were part of Art COP21, so we were approved artists, and that gave us some leeway.

The Climate Guardians have been a provocative presence in the opposition to the continued development of fossil fuel projects. Their practice continues even as there remains little restraint on the opening and expansion of new gas wells and coal mines for international export markets (Figure 16.5). It may be up to history to recall the work of the angels over more than a decade of activism, to retrospectively ask how Australia continued with its fossil fuel economy. In those future accounts, the appeal and enduring aspects of angels might just encourage us to hold on to a thread of hope.

References

AAP. 2021. 'Victoria Blocks AGL's Gas Terminal on Environmental Grounds.' *The Guardian*, 30 March. https://www.theguardian.com/business/2021/mar/30/victoria-blocks-agl-energy-gas-terminal-crib-point-split-business, accessed 15 May 2025.

Varney, Denise. 2019. 'Host of Angels: Climate Guardians and Quiet Activism.' In Peter Eckersall and Helena Grehan, eds, *The Routledge Companion to Theatre and Politics*. London: Routledge, 67–71.

17

WILDERNESS ACTS

Dorothy Molter and Performing the Boundary Waters

Eero Laine

We had been paddling our canoe for a few hours, each stroke of the paddle taking us further past the imaginary line that demarcated the edge of the wilderness. Beyond that border, there were no roads, no wayfinding signs, no motors, no electricity, no cell service. It was a hot day for the beginning of September in northern Minnesota, perhaps another reminder of the climate crisis and the ways it reaches all parts of the planet. We didn't do ourselves any favors in the heat as we weren't particularly packing light. We had two good sized canoe packs, one made of heavy grade nylon with padded straps and the other of a more traditional construction: waxed canvas and leather. They held a tent and sleeping bags, a two-burner stove, pots and a pan, clothes, ropes, tarps and various other odds and ends of camping gear. And we also had a sixty-liter waterproof plastic barrel full of food including a small cooler meant to keep fresh items from spoiling in the heat. The lakes we paddled were connected to each other by portages—rough trails through the woods connecting various bodies of water or offering safer passage around rough parts of rivers and streams When paddling, the canoe carried all of this without trouble. When crossing the portages, we carried it all with considerably less ease.

We finished our last portage for the day and arrived on Knife Lake. We were part of a stream of canoeists that then dispersed into the larger body of water, headed further east, perhaps to find a place to set up camp on the lake, perhaps with more days of traveling ahead, towards Lake Superior. We were ready to find a campsite. By most measures, we were in the middle of nowhere, hours of paddling and portaging from the nearest cabin or car or store. If I had been in that exact spot forty years earlier, however, I could have

DOI: 10.4324/9781003584445-22

paddled straight out from the portage and ordered an ice-cold root beer on the Isle of Pines, Dorothy Molter's home in the wilderness.

When she died in 1986, Molter was the last non-Indigenous permanent inhabitant of the Boundary Waters Canoe Area Wilderness. She played a singular role in the area and local communities, and her life is remembered and reperformed today through theatre, a museum, books and a line of root beer. Molter moved to Knife Lake before it was part of the designated wilderness of the Boundary Waters, and unlike the trip just described, Molter would have arrived by float plane or with motorboat—at least initially. Over the decades, she witnessed the wilderness take over, one legislative act at a time, as planes were outlawed, then private property was aggressively purchased by the state, and finally, all motorised vehicles were banned. This chapter examines the wild dramaturgies of wilderness and considers the life of Dorothy Molter as an example of an enduring wilderness performance. It begins with an overview of the prevailing definition of wilderness in the United States, its application to a particular wilderness area in northern Minnesota, and then examines Dorothy Molter's place and performance in that wilderness.

Wilderness Acts

The wilderness, of course, is a site of performance. Or it is as much a site of performance as any other site. But what perhaps makes it noteworthy is the way wilderness is so often defined as separate from or outside of those places where we might more regularly think about performance—cities, centres of culture, arts organisations and human endeavours, broadly construed. In some ways, wilderness performs because of its lack of performance. In living her life in the wilderness, Molter can be seen as an exception within a space of exception—a lingering reminder of the world that the wilderness recaptured but is nevertheless surrounded by today. That is, we understand wilderness as wild only in comparison to what it is not. Much like nature itself, which can be viewed as an invention of the past few hundred years (Wulf 2015), wilderness is constructed and performed by its very identification and definition.

The US Wilderness Act offers a legal definition of wilderness that has proven influential in various parts of the world following its passing in 1964. A key part of the act reads:

> (c) A wilderness, in contrast with those areas where man [sic] and his works dominate the landscape, is hereby recognized as an area where the earth and its community of life are untrammeled by man, where man himself is a visitor who does not remain. An area of wilderness is further defined to mean in this Act an area of undeveloped Federal land retaining its primeval character

> and influence, without permanent improvements or human habitation, which is protected and managed so as to preserve its natural conditions and which:
>
> 1 generally appears to have been affected primarily by the forces of nature, with the imprint of man's work substantially unnoticeable;
> 2 has outstanding opportunities for solitude or a primitive and unconfined type of recreation;
> 3 has at least five thousand acres of land or is of sufficient size as to make practicable its preservation and use in an unimpaired condition; and
> 4 may also contain ecological, geological, or other features of scientific, educational, scenic, or historical value. (US Wilderness Act 1964)

Few researchers in theatre and performance studies have examined wilderness as performance. Eric Wiley, however, has observed many of the legal features of wilderness, examining the performance of swamp tours near New Orleans that appealed to tourists by theming the swampland as a wild space of nature:

> The "wilderness" theme defines the swamplands in fundamental opposition to civilization. Added to this are the secondary qualifiers: the swamps are non-civilized, non-socially constructed, ahistorical, essentialist "nature," and are characterized variously by virginity, beauty, hostility, mystique, amorality, and timelessness.
>
> *(2002, 120)*

That is, these seemingly inherent aspects of wilderness are laid over the space and performed. The wilderness comes into being through tours, experiences and other interactions that frame them as wild. The wilderness is wilderness because it is twice behaved, perhaps. It becomes wilderness 'the second to the nth time' it is encountered as wilderness (Schechner 1985, 36).

Per the US Wilderness Act, the wilderness '*appears* to have been affected primarily by the forces of nature' (1964, emphasis added). Important here for theatre and performance studies, the language of the legislation does not say that the wilderness is only affected by forces of nature, but rather that it appears that way. This, to me, implies a staging of wilderness, where human planning and impact and preservation occur but only in a sort of backstage space that does not impact the natural appearance of the wilderness area. Thus, there is a dramaturgy to what one experiences in the wilderness. In some areas, this might be as simple as establishing a border and then keeping everything out and letting nature run its course. But the example of the Boundary Waters Canoe Area Wilderness (BWCAW) is a useful reminder that it takes immense efforts to establish the appearance of wilderness and even further efforts to maintain it.

The wilderness was not always the wilderness as we legally define it today, and perhaps it will not be the wilderness as we currently understand it into the future. A common refrain is that the waterways that span the US/Canadian border have not really changed, that they are somehow outside of time, that one travels through them in much the same ways that humans did hundreds or thousands of years ago. Yet, those people also shaped the area and left behind traces and evidence of their lives, for instance, various pictographs throughout the area. Some of these pictographs resemble the cave painting depictions of hands from tens of thousands of years ago described by Rebecca Schneider as 'Held up in and as stone, the hands are gesturing greeting, hailing, beckoning' (2018, 290). The hands and the canoes depicted in the pictographs are best viewed from the level of the lake and while in a canoe. They perform across eras and invite alternate histories and speculative futures to be written through and across those seemingly untouched areas of wilderness. They stage an encounter across time and remind us that people in the areas have come and gone for a very long time.

Wilderness performances are unstable and slip by quickly in the long span of geologic time. Yet, sifting through the layers of more recent legislative strata, the work to create or curate a wilderness can take a lifetime. In the past century, for instance, not only has quite a bit changed, but also this area, seemingly today out on the periphery, wild, natural, has actually been at the centre of national political debates and fights, none of which were in any way obvious or uncontested in their times. The wilderness doesn't simply exist because it was overlooked. It exists because people spent their lives defining it, fighting for it, and defending it from enclosure. And in the case of the Boundary Waters, dispelling private interests and ownership—essentially un-enclosing a common area. That is, far from being 'pristine, untouched wilderness' as it is often described, the Boundary Waters area is wilderness because it has been made to perform as wilderness, through legal battles, legislative decisions, and through representations that have shaped its imaginary pasts and futures.

On the US side of the border route, nineteenth-century treaties with the Ojibwe nation emboldened the United States to continue resource extraction in earnest. By the late 1800s, the area was being extensively mined and logged, parcelled out to buyers, both local and international. That slowed somewhat as a large area of northern Minnesota was designated part of the US National Forest, both an early conservationist effort and an extension of the colonial project that named land as federally owned and operated. By the late 1920s, areas were designated as part of a roadless zone and then as the Superior Roadless Primitive Area, which effectively ended commercial development and logging as no new roads were allowed. Many businesses in the area pivoted towards recreation as resorts and fishing and hunting lodges cropped up and expanded throughout the region.

The BWCAW is an area that was compelled to perform wilderness as the space of the non-wilderness receded—businesses, residences, infrastructure specifically removed, peeled back and pushed out. This campaign to remove the outward evidence of human impact is central to the US Wilderness Act, and as the act was deliberated, now decades ago, much attention was paid to the affective impact of how the wilderness felt: 'Sensory references to sight, sounds, and smells were commonly used to represent civilization and cities as chaotic, confusing, and unnatural. Wilderness advocates argued for the necessity of "solitude" in wilderness to escape the unnatural sensory overload of modernity' (Engebretson and Hall 2019). The absence of the sounds of cars, motorised boats, float planes, and other modern machinery is a dramaturgical choice in establishing the wilderness.

Dorothy Molter's Wilderness Performance

There is an island in northern Minnesota called the Isle of Pines. It is where Dorothy Molter lived. It is a small island and it sits on a remote lake. That lake may have been home to Paleoindians over 10,000 years ago, and it is more commonly referred to in Ojibwe as *Mookomaan Zaaga'igan*, *Lac des Couteaux* in French or Knife Lake on many maps today (Muñiz 2013; Unze 2011). Knife Lake is part of a chain of lakes that make up the contemporary US/Canadian border, the 'border route' for recreational paddlers and campers or the 'ancient superhighway' as *National Geographic* has called it, earning the title as one of 'the most important Indigenous transportation routes in North America' (Pearson 2022). The French Voyagers each would have hauled heavy bales of beaver and otter fur along this route, stealing them away to colonial ships in Lake Superior before returning to trapping and trade routes that extended west across the US and Canadian lakes and woods and north to Hudson Bay. The terrain is rugged, traversing across windswept lakes and rushing rapids and muck-filled streams. Canoes are indigenous to the region—symbols and essential tools of those who lived in the area for centuries before they facilitated the concomitant import of European settlers and the extraction of animal hides, timber and precious metals (Erickson and Krotz 2021). Today, canoes remain the only way to navigate the waterways of the Boundary Waters Canoe Area Wilderness.

And that is how Dorothy Molter, the owner and inhabitant of the Isle of Pines, travelled to and from her home on Knife Lake for nearly six decades, from when she moved there shortly after finishing nursing school until her death in 1986. My early September trip to Knife Lake, described at the beginning of this chapter, was Molter's commute between her island home in the wilderness and the outside world. Today, she is remembered in the nearby town of Ely, Minnesota as the Root Beer Lady because she made and sold root beer to passing canoeists. She has her own museum in town (Huyck 2020), a root beer for sale at the local grocery store, and at least two plays dedicated to her life (Laine 2024; Schultz 2023). Her obituary in the *New York Times* reminded readers that she had once been described as the loneliest woman in America, living alone on her island in the wooded wilderness of northern Minnesota (1986).

Perhaps Molter really wasn't that lonely on her island—or at least not at first. She certainly had many visitors and hosted friends throughout the year. Until the middle of the century, she lived on the island with Bill Berglund, the original owner of the Isle of Pines resort. Anglers and adventurers would charter float planes, fly into the now beautifully regrown roadless area that had not been logged for decades, land on a lake, check into a resort like the one in which Dorothy Molter lived and worked on her island and fish or hunt for a week or more. Berglund died in 1948 and the planes were banned in 1949. Then, after decades of debate and activism, the Wilderness Act of 1964 solidified the Boundary Waters Canoe Area in northern Minnesota as a protected wilderness, but she still had many visitors by canoe in the summer and snow machines in the winter. In 1979, motorised travel was banned from the area and the winters were much quieter.

Amidst all of this, Dorothy Molter, eventually alone on her island, refused to give up her property, instead selling snacks and supplies to passing canoeists. When the float planes and motors were banned, which interrupted her supply chain, she reused the empty bottles on her island to bottle root beer that she could brew with lake water and dry goods that were more easily brought in by canoe. She offered this root beer to thirsty campers in the summer, for a donation in order to get around prohibitions on commerce in the wilderness. Over the course of her life, Molter saw modern conveniences and connections to society recede, one congressional bill or executive order at a time. People who had been neighbours across the lake now visited only by canoe in the summer or by snow machine and then by dogsled in the winter. She inhabited the wilderness in solitude, eventually known to those who passed by as the Root Beer Lady. This durational performance in the wilderness places Dorothy Molter and the ways she is remembered in a rather unique position. On the one hand, she is deeply imbricated with the woods in which she lived, a symbol of the wilderness spirit, an independent woman fully untethered from social dictum. On the other hand, she is the last holdout in what some at the time saw as government overreach or a clear encroachment on private land and free commercial activity and others thought of as the path toward preservation and ecological stewardship. This fraught dynamic plays out today in the nearby town of Ely, Minnesota.

Wilderness Legacies at the End of the Road

With a current population of just over 3,000 people, Ely is frequently referred to as both the 'Canoe Capital of the World' and 'the end of the road'. The town is the last stop before one literally runs out of road at the edges of the wilderness, and then, the only way to travel is by canoe or by foot. It is an identity on the border of wilderness that is embraced by the community. There is an annual canoe portage marathon, the local radio station WELY is referred to as 'End of the Road Radio', and if global climate change might indicate a figurative end of the road, the Boundary Waters has become a vital front for national discussions related to climate change and preservation, especially as devastating wildfires and wind events become increasingly common.

But the threats to the area are not only at the level of global catastrophic warming. A copper-sulphide mining operation in the Boundary Waters watershed is hotly debated in local and national politics. Presidential holds and approvals have swung back and forth from the Obama administration to Trump, mining was delayed by an executive order from Joe Biden (Kraker 2023), and the conservative Project 2025 targets the area for deregulation (Kraus 2024). Lawn signs in Ely plead with visitors and locals to 'Save the Boundary Waters', while others let passersby know that the residents of the house 'support mining'. It is a touchy subject in the small town, and the lines are bitterly divided. But that is not new. When the US Wilderness Act passed in 1964, some who opposed the act hung an effigy of conservationist Sigurd Olson in the streets of Ely, an event people will still mention today in conversation. Prior to that, float plane companies and pilots, many of whom were US World War II veterans, vehemently opposed presidential actions that limited and then eliminated their business in the area. And it is perhaps not difficult to imagine the outrage of resort and private landowners in the area as they were compelled to sell to the government, their family cabins and business properties then burned or sunk to the bottom of a lake, erased in order to establish or perhaps re-reperform wilderness.

Today, well beyond the end of the road—on Knife Lake, on the border route, in the Boundary Waters Canoe Area Wilderness, the Isle of Pines is empty. The forest has begun its slow creep back over sites that used to be cabin platforms, an ice shack and boat storage. The trails on the island remain walkable but only due to visitors who come to see where Dorothy lived, to imagine a life in the wilderness, to reperform the steps that Dorothy Molter would have taken for over half a century, now another third of a century since she took her last steps. The trails will only exist as long as they are used. Still, when I was there, trees had blown or fallen down across some of the trails, making them impassible. Perhaps this is one small example of the wilderness, rewilding, reperforming itself again and not for the last time.

It is often difficult to believe that many lakes in the region had thriving businesses, regular float plane deliveries, arrivals and departures. That was now a lifetime ago. One might consider a speculative past marked by the Isle of Pines, hailed by Dorothy Molter's regular reperformance across stage and museum and root beer bottles. In these reperformances and while walking on Dorothy's island, perhaps like the pictographs, we might imagine an alternate present and future shaped by another history. It is not too difficult to consider what would be without the Wilderness Act or other presidential or congressional interventions. It is a past that leads to our present, wherein Molter's Isle of Pines is just one privately owned island among many in the Boundary Waters. Like most other places in the United States, private property would be bought and sold, developed and razed. Tourism and attractions would have expanded unchecked beyond fishing and hunting to family entertainment and attractions—concert series and waterparks and weddings and holiday events. Certainly, these are familiar in many other wooded areas with lakefront views, long since capitalised upon. Why wouldn't a hotel chain build a location on the Isle of Pines? Or imagine Dorothy expands her candy and soda business, a pizzeria opens an island over, and a sandwich and beer shop across the shore, wouldn't it seem inevitable that a Starbucks would move in at some point during this speculative timeline?

Now, imagine the Wilderness Act is rather passed today, not 60 years ago. The pizzeria, the hotel, the bar and café and the Starbucks are forced to sell, the family cabins that dot the shores of the lake are bought and burned by the forest service, the pontoons and waterski courses are sunk to the bottoms of the lakes, all signs removed. In a few decades, we might stand on that piece of shore with some remaining trails where people used to queue in line for their coffee and say 'There used to be a Starbucks here' much the same way today one can visit the Isle of Pines and say, there used to be a resort here or Dorothy Molter used to live here, she walked this trail. She fed the ducks and made root beer. Everything done can be undone. The wilderness exists because we perform it. It is something one can't help but think while paddling past the now-empty Isle of Pines. It is a reminder for us, at what feels like the end of the road, that everything could have gone and still can go another way entirely.

References

Engebretson, Jesse M. and Troy E. Hall. 2019. 'The Historical Meaning of 'Outstanding Opportunities for Solitude or a Primitive and Unconfined Type of Recreation' in the Wilderness Act of 1964.' *International Journal of Wilderness* 25 (2). https://ijw.org/historical-meaning-in-wilderness-act/.

Erickson, Bruce and Srah Wylie Krotz, eds. 2021. *The Politics of the Canoe*. Winnipeg: University of Manitoba Press.

Huyck, Heather. 2020. *Doing Women's History in Public a Handbook for Interpretation at Museums and Historic Sites*. Lanham, MD: Rowman & Littlefield.

Kraker, Dan. 2023. 'Feds Slap 20-Year Mining Ban on Land Near Boundary Waters.' *MPRNews*. 26 January. https://www.mprnews.org/story/2023/01/26/feds-slap-20year-mining-ban-on-land-near-boundary-waters, accessed 19 April 2025.

Kraus, Louis. 2024. 'Project 2025 Platform Proposal Aims to Allow Mining in Boundary Waters Watershed.' *Minnesota Star Tribune*, 7 November. https://www.startribune.com/project-2025-platform-proposal-aims-to-allow-mining-in-boundary-waters-watershed/600385719, accessed 19 April 2025.

Laine, Eero. 2024. 'Review of *Root Beer Lady the Musical*, by Barbara Cary Hall.' *Theatre Journal* 76 (2), 241–42.
Muñiz, Mark P. 2013. 'Exploring the Paleoindian Occupation of Knife Lake, Superior National Forest, Minnesota.' *Minnesota Archaeologist* 72, 113–57.
New York Times. 1986 'Nation's "Loneliest Woman," Dorothy Molter, Is Dead at 79.' 20 December, 34.
Pearson, Stephanie. 2022. 'Paddling Minnesota's "Ancient Superhighway".' *National Geographic*, 27 April 27. https://www.nationalgeographic.com/travel/article/paddling-minnesota-ancient-superhighway-border-route, accessed 19 April 2025.
Schechner, Richard. 1985. *Between Theater and Anthropology*. Philadelphia: University of Pennsylvania Press.
Schneider, Rebecca. 2018. 'That the Past May Yet Have Another Future: Gesture in the Times of Hands Up.' *Theatre Journal* 70 (3), 285–306. https://doi.org/10.1353/tj.2018.0056.
Schultz, Kim. 2023. 'The Root Beer Lady.' *National New Play Network New Play Exchange*. https://newplayexchange.org/script/2046759/the-root-beer-lady, accessed 19 April 2025.
Unze, David. 2011. 'New Info on State's Earliest Inhabitants Found in BWCA.' *MPRNews*, 2 October. https://www.mprnews.org/story/2011/10/02/bwca-research, accessed 19 April 2025.
US Wilderness Act. 1964. 'National Wilderness Preservation System,' 16 USC 1131–1136. *United States Code*, 2018 ed., Supplement 5, Title 16. https://www.govinfo.gov/app/details/USCODE-2023-title16/USCODE-2023-title16-chap23-sec1131/summary, accessed 19 April 2025.
Wiley, Eric. 2002. 'Wilderness Theatre: Environmental Tourism and Cajun Swamp Tours.' *TDR: The Drama Review* 46 (3), 118–31.
Wulf, Andrea. 2015. *The Invention of Nature: Alexander Von Humboldt's New World*. New York: Knopf Doubleday Publishing.

Culture, Community and Survival

18

HYPHENATED THINKING IN PERFORMANCE-PEDAGOGICAL PROTOTYPES

Common Dreams Flotation School and Natural Contract Lab

Christel Stalpaert[1]

In his books *A Decolonial Ecology* (2022) and *Loving Ourselves the Earth* (2024), Caribbean environmental engineer Malcom Ferdinand explores the relation between the current ecological crisis and the colonial history of modernity. He convincingly argues that the economisation of nature bears with it a history of colonialism. In *The Nutmeg's Curse: Parables for a Planet in Crisis*, Indian writer Amitav Ghosh explains how colonialism is intertwined with Western Enlightenment thinking. According to Ghosh, philosophers like René Descartes promoted the triumph of modern, progressive rationality (2021, 89). The mechanistic worldview of the Western Enlightenment focusses on the 'superior' human being who excels in the ability to speak clearly and with persuasion, and who progresses in life by acting purposefully and efficiently. In this respect, nature is primarily a repository of resources that help human progress. Ghosh observes how the 'mechanistic vision of the world' proclaims not only a process of progress, expansion, conquest and extraction but also of segregation: the so-called 'civilised' elite could distinguish themselves from the 'primitive' peoples on the basis of their belief that rational knowledge production was the only valid one (2021, 37). Western thought considers the cognitive development of 'common sense' still superior to other forms of knowledge production, such as embodied or intergenerational knowledge. Ghosh calls upon an ethics for decolonialisation both in ecological and pedagogical terms.

Despite the noble ideals of educational institutions (at least after the First World War)—that is, to grant democratic access to citizens across classes and beyond privileges, to increase literacy and gender equality worldwide—we are still struggling with the persistent relics of a mechanistic-imperialist worldview in the current landscape of infrastructures of knowledge. Epistemic violence is

DOI: 10.4324/9781003584445-24

still at work. Moreover, since the 1980s, progressive rationality took a leap, cultivating an obsessive focus on the efficient production of knowledge and pushing research into competitive extremes. Within what philosopher and art theorist Gerald Raunig calls the 'machinic thinking' of 'cognitive capitalism' (2013, 17), the Western machine of knowledge transfer is increasingly driven by capitalist valorisation. In *Factories of Knowledge* (2013), he describes how knowledge is manufactured as a material commodity, accessible only through an ever-increasing (registration) cost, and traded within the competitive logic of hierarchically ordered publication venues and funding parameters. Knowledge is

> passed on from one to another as property, from teacher to pupil in a long chain of tradition, in a hierarchy of generations and a uniform, static order of knowledge. In this order, the various techniques and disciplines are rigidly striated and separated.
>
> *(2013, 56)*

In this machinic thinking, students are merely 'passive components of the knowledge factory, as formatted knowledge *re*producers' (2013, 42). Moreover, this knowledge factory encourages segregation. Western thought considers the cognitive development of 'common sense' still superior to other forms of knowledge production, which are situated, embodied, or entangled. Raunig argues for disobedience: we may be cogs in a knowledge machine, but we can also sabotage the flow of knowledge production. According to him, there is room for social struggle and resistance, just as there was in the classical factory. 'We have to develop a risky practice that seeks conflict and transversal exchange beyond the boundaries of traditions and disciplines', he says (2013, 56).

From my own positionality within the Arts Studies department at Ghent University, I am constantly asking myself how we can address and challenge the relics of the mechanistic-imperialist worldview in our education system. How can we prevent the mesmerisingly fast neoliberal machine and its symbols of power from driving us into a slumber mode? How can we move away from the cognitive-capitalist timescale, where 'hurry is the normality' (Raunig 2013, 84), and ensure that ideas are given the necessary time to mature? How do we create a collaborative learning environment to share that time? How can we stop segregation and epistemic violence in the rational-progressive flow of knowledge production? Does the obstructive strategy of sabotage suffice in this regard?

Sabotage with Bouncing Effect in Machineries of Knowledge Production

Sabotage, as a form of resistance, is usually described in negative terms: it is considered a malicious opposition, a wilful damage, a hostile intervention and a blockade. 'Sabotage' is usually defined by its obstructive power: by renouncing to do something, by stopping or slowing down the production process, etc. They are forms of political resistance or civil disobedience that manifest themselves by causing the course of events or a production process to fail or impede. This negative connotation is related to the origin of the word. The *sabot* is the French word for clog and refers to the wooden footwear worn by factory workers in the second half of the nineteenth century. They were an ideal means of poking between the wheels of the factory machine and halting the flow of production. Initially, this was done in case of emergencies: the large, heavy clogs were strong enough to make the cast-iron machines stop suddenly and prevent an accident. But this strategy was soon adopted as an act of resistance. Flemish workers in Liège are said to have been the first to put their clogs into the new machines during the Jacquerie strike of 1886 to express their displeasure with the poor and suppressive working conditions in their factory.

However, sabotage can be more than a mere obstructive force. It can be both an impeding and transformative force in the rational-progressive view on production processes. This is because when the flow of movement in the production process is disrupted, energy is released. In physics, an obstruction can lead to shape change or heat production, a frictional force, a new shear force or deceleration. It causes worn surfaces, as well as polished or brushed surfaces. At best, the whole machinery is shaken up and a stick-slip effect is created: a bouncing effect that causes the machine to start creaking, scuffing and whining. A spontaneous jerking motion is then created that is difficult to control.

Imagine the opportunities that bouncing sabotage strategies could create in the rational-progressive machineries of knowledge production, such as the university. Think of the opportunities for transforming (and not only obstructing) the persistent relics of a mechanistic-imperialist worldview into an ecology of care. Just imagine the energy that is released when we sabotage individualistic, hierarchical knowledge transfer. When we create shared and collaborative learning environments that transcend hierarchical didactics. What transformation or change of epistemological framework occurs when artistic and performative modes of knowledge production are welcomed in an academic university context? What time regime is released when sabotage slows down capitalist valorisation? How does the university machine of cognitive-capitalist knowledge production transform when it is bouncing of creativity and diversity, while unlearning the process of segregation and epistemic violence?

In order to have our university bounce of creativity and diversity, it will take more than the occasional counterexample during a lecture to supplement or correct the canon. Epistemic violence does not stop with providing an occasional guest lecture. In these cases, the 'other' is all too easily treated as a *sabot* or clog in the iron machine of knowledge production, without anyone within the machinery taking up a responsibility themselves to dismantle the machinery.

And, there is a high risk that the clog—and not the machinery of knowledge production—breaks under so much counterweight. More than being a mere vehicle for critique or a form of deconstruction, decolonial theories and practices involve an engagement with unlearning, undoing, recasting and transforming Eurocentric knowledge formations. For filmmaker and cultural scholar Ariella Aïsha Azoulay, activating such an epistemological change means unlearning imperialism. Inspired by philosopher and activist Audre Lorde, Azoulay proposes to disable the master's tools of white, patriarchal knowledge institutions. Thinking with Lorde and Azoulay, this contribution then wonders how we can sabotage the institutionally embedded, mechanistic-imperialist mindsets, transforming 'non-qualified' forms of knowledge production into 'one of many' (Azoulay 2019, 385).

'Unlearning in a series of rehearsals (...) with the help of other companions' (Azoulay 2019, 15, 29)

Focussing on the university as an infrastructure of knowledge, a key question then is how to be employed in the university factory, without silently moving along with the machinery. Azoulay questioned her own position at Brown University, one of the eight US so-called Ivy League universities. For her, the challenge was not being seduced by the 'accelerated project of the neoliberal American university', not becoming the ambassador of imperialism, and not normalising the privileged access to research objects offered to scholars. I read Azoulay's book as an invitation to keep sabotaging the knowledge machine of which I am part of (but refuse to belong to). In a series of rehearsals, and with the help of many companions inside and outside the university, I try to trigger the bouncing sabotage effect in my university as a factory of knowledge and did so over the past decades. Call me a saboteur who functions by the grace of the connecting hyphens with artists, scientists, activists and students.

The Ghent-based, Portuguese eco artist and water steward Maria Lucia Cruz Correia is one of these long-term companions. Correia initiated the first *Common Dreams Flotation School* in Ghent in 2017. This resulted in a survival climate school built on water co-creating and (re)collecting strategies for survival in a floating society (see also Stalpaert 2023). The School was co-created with university college KASK, Art Centre VIERNULVIER, the city of Ghent, MEDS Network (Meeting of Design Students) and several local stakeholders, such as Jonge Dokken (representing the community of boat inhabitants), Buurderij (a local network of farmers and consumers) and Thuishaven (a local service centre), each having their particular stake in the neighbourhood of the river Lys, the river Scheldt and the harbour of Ghent. An important question of the school was related to finding new tools in replacement of the master's tools: 'what kind of tools do we need as a gesture of care for the planetary?' (Correia 2025).

In 2021, Correia initiated the Natural Contract Lab (NCL) in collaboration with Vinny Jones (sensory scenography/dramaturgy), Lode Vranken (design/philosophy), Brunilda Pali (restorative justice), Margarida Mendes (research/sonic guidance) and SZenne River: a transdisciplinary, co-creative group that brings together art, environmental law, restorative justice, rights of nature and ecological grief.[2] Its artistic practices aim at reconsidering our relation with bodies of water. Throughout the years, the Lab developed an artistic *Protocol of Reciprocal Care* to imagine new forms of environmental justice and to realign the harm inflicted on the hydro commons.

I joined the NCL team for *STILL HERE: An Alliance of Care for the SZenne River* (2023–2025), together with many other artists, scientists and stakeholders of SZenne River (see also Stalpaert 2025). This performance project takes place across the three regional territories of Flanders, Wallonia and Brussels Capital—from the source of SZenne River in Soignies, to its bifurcation at the Zennegat in Mechelen. Artists include Matteo Deblasio (river vessel/clay mediator) and Marzia Dalfini (walking-with costumography). Scientists include River Guardian pedagogue Saartje Monden and environmental lawyer Jef Seghers. I am engaged as 'a hyphenated thinker' and assist in developing tools for tracing and transferring the repertoire of care for SZenne River.[3] The term 'stakeholders' refers to the (multispecies) beings that have a direct interest in the hydro-ecology of SZenne River. These range from local inhabitants, such as residents along SZenne River, its beings, waters and cosmology, to environmental stakeholders monitoring the area, such as the Walloon organisation Contrat de Rivière Senne, Coordinatie Zenne, Brussels Environment, Natuurpunt Mechelen and Zenne Garden Collective. In the autumn of 2025, the artistic practices developed by NCL with SZenne River are implemented as a 12-week research seminar on Art & Ecology in the Arts Department at Ghent University, seeking a 'transmission of knowledge and sharing of our protocol of reciprocal care' (Correia 2025).

Hyphenated Thinking

Like New Materialist thinker Donna Haraway, I believe in the world-making potential of a particular coalition of art, science and activism in the struggle for a partial resurgence in this troubled world. Building on her concept of 'art-science-activist worldings' as a way for 'staying with the trouble' (2016, 79, 76), I coined the term hyphenated thinking: a method of thinking with these long durational collaborative art practices in which the discrete boundaries between disciplines, fields of study and research domains are permeable (Stalpaert 2020).[4] Implementing these art practices in a university context means moving-with NCL in-between art, science and activism. This interdisciplinary moving-with sabotages the compartmentalised thinking in faculties, departments, disciplines, specialised research areas, research groups and modes of knowledge production in universities. As the roles of artists, scientists and stakeholders-cum-activists are well-defined in NCL, the implementation of their fields of study in the group seems logical. However, the moment anyone joins NCL, they join the hyphenated alliance of care for SZenne River. This means that scientists do not deliver matters of fact from their field of study to support the representation of a better world. Everyone involved cultivates matters of concern and care and fabulates a common world-to-come in-between art, science and activism. They all participate in site-responsive artistic walkings-with SZenne River and sensorial mappings, they all rehearse the storying of (hydro-)ecological matters of concern and care through conversational circles, they all exchange water memories, rivers stories and ecological grief testimonies. And they all rehearse a weaving together of individual grief memories of the landscapes they have lost, 'as a gesture of solidarity for the imagination of a sensory preamble to propose into a Living Bill for SZenne River's legal rights' (Correia 2025). Besides their full engagement throughout the process, the alliance also collaborates on joint articles, performance lectures and work demonstrations. The *wheres* and *hows* of these collaborations are shaped by the needs that emerge over time, and in the case of *STILL HERE*, thinking is constantly hyphenated through the shared laboratory culture.

Taking seriously decolonial urges, I consider such a hyphenated thinking also an opportunity for unlearning mechanistic-imperialist and cognitive-capitalist worldviews. Common Dreams Flotation School and NCL transform the accustomed format of academic knowledge production organised around one privileged (cognitive) centre and along a hierarchical organisation of knowledge transfer. In any case, the hyphen validates an in-between zone for pluriversal, multispecies and collaborative thinking. Raunig calls it a relational imparting of knowledge: a 'moving along a relationship (or multiple posited relationships) without fixing the production of knowledge in a firm center' (2013, 58). The artistic practices in Common Dreams School and NCL generate situated, embodied, and entangled modes of knowledge production. Embedding the artistic practices in the interdisciplinary setting of a university infrastructure means transforming these 'non-qualified' forms of knowledge production into 'one of many' (Azoulay 2019, 385). It is this hyphenated thinking that I consider valuable, reflecting on the climate-related challenges that our world is currently facing.

A hyphenated thinking is not an applicable method; it is foremost an open learning and research attitude, in this case, activated by particular art practices. Haraway's concept of 'worlding' refers to an understanding of the world as a verb. She focusses on practices instead of representations of social and environmental relations. A hyphenated thinking is for that matter a world-making practice. It is an enacted and embodied activity and hence implies performativity. It requires other forms of attention and language than our science-accustomed. The enacted practice of a hyphenated thinking in Common Dreams

Flotation School and NCL is a speculative fabulation, a continuous exploration of our radically enmeshed becoming-*with* and attending-*to* a world-to-come. We unlearn mechanistic-imperialist modes of thinking and practice alternative forms of knowledge production in a series of deep ecological rehearsals of care for water bodies. Common Dreams Flotation School and NCL are world-making practices, but in a sense that humans are no longer understood as individual subjects that can know, control, exhaust and destroy their environment. The performative zone in what I call performance-pedagogical prototypes proves to be very fruitful for acknowledging our becoming *with* the world. Rather than being a study of our being *in* and control *of* and *over* the world, the performative zone cultivates an attitude that defies human exceptionalism and entitlement.

Performance-Pedagogical Prototypes

With every edition, the School or the Lab evolves differently, depending on the particular location, the engaged knowledge institutions and the particular stakeholders involved.

Common Dreams Flotation School, for example, is built on water and (re)collects strategies for survival in a floating society. The neighborliness with water in this site-specific school constantly echoes current climate challenges such as rising sea levels, heavy rainfall and extreme flooding. The School has participants experience how the global population living in coastal areas or river basins are at risk. The first edition of the Common Dreams School was co-created over the course of the academic year 2017–18 and took the form of a floating raft on the river. In that context, Correia calls the school 'a mobile prototype' of a survival climate school (Correia n.d.). It is mobile, as the school is swiftly adaptable. It can travel to other cities and can easily adjust to other collaborative alliances with local universities, art institutions and sustainable organisations. It is a prototype, as the survival climate school deliberately remains an early sample and its productivity is rather limited—as far as measurable output and impact is concerned. The Common Dreams Flotation School is too situative, too delicate and too speculative to ever be considered for mass production.

One could say that Common Dreams Flotation School is a continuous construction site and has no intention or desire to become a pedagogical institution. With every new location that the School is residing, it starts from scratch. It is literally constructed from material found in the neighbourhood, and in every new edition, the construction site evolves differently. It can take the form of a floating raft in a harbour, or on a river (Ghent, 2017), or on a lake (Geneva, 2020); in a tent in the forest (Geneva, 2020); or it can simply take the form of interventions in a public space (Brussels, 2020).

Collaborating with the different stakeholders in the area, Common Dreams School also considers space not as something that is given or owned, but as something that is constantly negotiated. The school is, in that sense, also a construction site in a figurative sense, in a way that it attributes qualities of undecidedness to a particular rendering of space:

> When you are guided to any construction site you are experiencing the troubling and exhilarating feeling that things *could be different*, or at least that they could still fail — a feeling never so deep when faced with the final product, no matter how beautiful or impressive it may be.
>
> *(Latour 2005, 89)*

In the subsequent Natural Contract Labs, the continuous negotiation of space is connected with the situated knowledge of a particular river area. As such, the multispecies alliances of NCL with Sijoumi Lake in Tunisia (2021–23), Rivers Tejo in Portugal (2021–23) and Rhone in Switzerland (2022–23) constituted very different processes. By doing so, NCL affirms the (usually underexplored) dynamic force of water-as-matter and the entanglement of water bodies, knowledge, and agency. Also, the diverse range of artists, scientists and stakeholders-cum-activists gathered in the particular area, generates different and sometimes conflicting perspectives on the (hydro-)ecological matters of concern and care that are raised. And this is the intention. The Lab does not aim to propose one solution to a problem, but to connect the participants in an alliance of care. This means first and foremost acknowledging the human entanglement in the particular water ecologies and the importance of safeguarding the well-being of its biotopes.

When these sustained art practices are implemented in an academic university context, it basically means that scientists and students are joining what I call a performance-pedagogical prototype (with a hyphen). These prototypes are no longer mere performances or pedagogical contexts, as they adopt elements from one another. The performance constitutes a particular pedagogical context, in the sense that it activates an unlearning and rehearsing, while the pedagogical university context gains performative ground when adopting its artistic practices as tools for knowledge production. Within the performative zone, the divide between audience, artwork and scholar is no longer tenable and gives way to a networked constellation. Every participant cannot help but deploy a hyphenated thinking in-between performance practice and pedagogical method. From the start, our habitual hierarchical position as student, teacher, artist, researcher, software developer, activist, spectator, performer, etc., shape-shifts throughout the twelve sessions. Our mindset is recalibrated repeatedly, as our thinking unfolds in an alliance of interconnectivity.

Bouncing Sabotage Effect Inside and Outside the University

Implementing a Common Dreams School or an NCL infuses a university infrastructure not only with a performative zone, but also with the bouncing effect of sabotage. The sustained engagement with artistic practices exceeds the guest lecture or workshop timespan of three hours. The Crafting at Dusk workshop in the Ghent Common Dreams Flotation School, for example, lasted for 12 hours. Together with curator and (performing) artist Ingrid Vranken, the participants camped on the floating island and speculated on how to survive on the water, while having an experience close to what life could be as a floating society. They were asked to bring food that would last on an island for a long time and that they could share with others. They could bring along water for one day, and one item that they would take from their home if there were a catastrophe, and present it during the workshop as survival gear equipment: a fishing gear, a knife, a mug, pen and notebook, a raincoat…. The idea is to fabulate strategies of survival from a situated, networked, intergenerational and embodied perspective:

> What stories will we leave for the future generations? How to create a shared intergenerational knowledge?
>
> *(Common Dreams Flotation School 2017)*

Likewise, the site-responsive artistic walking-with SZenne River takes a whole day and the long duration intensifies our response to the particular riverscape as networked environment. Exceeding the usual three-hour seminar format, the workshop sessions slow down time against the grain of clock-time, distributing and dividing it up differently than our science-accustomed ones. Also, when walking and workshop time is extended over several sessions throughout the academic year, private and public school time are blurred. Does the walking-with that is rehearsed on an individual basis belong to the private or public school time? In any case, the timescale differs from the capitalist valorisation of 'hurry is the normality' (Raunig 2013, 84) because of it stretched and sustained nature.

Because of its collaborative nature, the alliance for SZenne River will remain operative long after NCL has left the area. To put it in Haraway's terms, the Lab is 'cultivating conditions for ongoingness' (2016, 38). There is a continuous 'making and unmaking' of the performance, with an alliance of collaborators soliciting response, 'passing on and receiving' artistic practices, 'picking up and dropping' information (2016, 3). As such, *STILL HERE* is an artistic practice and an ongoing learning process at the same time. The sensorial mapping and the conversation circles, for example, constitute a repertoire for gestures of care for river bodies. During the Walking-with SZenne River, we unlearn cartographic mapping methodologies by following practical, yet poetic propositions and by rehearsing a deep sensorial mapping. Common Dreams School and NCL—continuously under construction—are in that sense also pedagogical prototypes that sabotage 'machinic thinking': they function as a site in which knowledge production is 're-invented' (Raunig 2013, 26). It functions as 'a complex space' where 'the most diverse forms of cognitive, affective, subservient labor' overlap (2013, 24).

The energy released through this sabotaging creates a precarious form of autonomy *within* a knowledge institution. The practice of schooling in NCL resonates for that matter with Ahmet Ögüt's Silent University (2012), Hamedine Kane & Stéphane Verlet-Bottéro's School of Mutants (2018) and Amanda Piña's School of the Jaguar (2019), amongst others. They *do* exactly this: they constitute 'a place where different forms of knowledge can enter into dialogue without the habitual hierarchies between them, exploring a variety of formats in which to operate, including the performative, the experiential, the discursive and the visual' (Piña 2019, 48). As an instituent practice, this practice of schooling sabotages univers(al)ity and activates plurivers(al)ity (Yakoub 2023).

The School or Lab becomes the university's free space for nonconformist thinking and action. The Radical Survival Tool Kit that evolves from the Crafting at Dusk workshop, or the co-authored preambule that feeds a law for the rights of SZenne River also sabotages the institutional competitive pressure on individual output in peer-reviewed and hierarchically ranked journals with a high impact factor. Its impact factor lies elsewhere. Rather than stimulating competition, it explores our capacities to act together, to imagine differential futures and common dreams, to develop survival tools together, and to rehearse neighbourgliness and kinship structures. In these turbulent times, it might be just these situated, embodied, and entangled knowledges that might be of future use.

Whereas Correia (2025) prefers to speak of Common Dreams School and NCL in terms of 'artistic fabulation, regenerative activism, or restorative justice', I also experience the artistic practice of schooling as a sabotaging practice with bouncing effect. Considered from my particular positionality at Ghent University, my invitation to NCL also entails an activation with sabotaging effect. The performative zones installed in the university's pedagogical context affect the mechanistic-imperialist master's tools in the current landscape of infrastructures of knowledge. They 'thwart the structures and institutional antagonisms through their obstinacy' (Raunig 2013, 27). Implementing the artistic practice of schooling within the academic university infrastructure instils the machinery of knowledge production with vibrant tools of critical pedagogy. Raunig would call this sabotaging practice 'a desertion as instituent practice' (2013, 27). Desertion here does not refer to military desertion, or to inner withdrawal and retreat from the world. On the contrary, it transforms a defection from the dead-end situation of the individualistic, competitive institutional practice of productive knowledge transfer into a collaborative alliance with many companions inside and outside the university. The risky practice of 'desertion as instituent practice' hence creates not only friction in disobedience but also a (little) more future-proof world by constituting an interdisciplinary alliance. After all, as Raunig claims, 'Desertion does not mean praise for fleeing from the world, but rather creating worlds' (2013, 53). Collaborating with local stakeholders and policy makers generates 'local and situative knowledge' (2013, 28) that feeds the acknowledgement of our ongoing worlding response-ability and that also concerns matters of entangled self-formation, self-control and self-care in times of crisis.

It is in this sense that the hyphenated thinking in performance-pedagogical prototypes cultivates 'conditions for ongoingness' (Haraway 2016, 38), for 'staying with the trouble' (76). To Haraway, staying with the trouble is 'learning to be truly present' (1) and 'getting on together' (10). 'The task is to become capable, with each other in all of our bumptious kinds, of response' (1). I agree with her that this ongoingness can only be acquired by emphasising connectedness and by de-emphasising human exceptionalism in favour of multispecism and pluriversity.

To be continued...

Notes

1 The performance project STILL HERE: An alliance of care for the SZenne River (2023–2025) is funded by the Flemish Government, Erfgoedcel Mechelen, Zenne Archives of Rijksarchief Brussel (BE), and the European Forum for Restorative Justice (EFRJ). The project facilitates a long-term commitment with Natural Contract Lab and an ephemeral performance process that is produced by HIROS, in co-production with VierNulVier, workspacebrussels, Kanal Centre Pompidou, Sint Lucas Antwerpen School of Arts KdG, and the Research Centre S:PAM (Studies in Performing Arts & Media) at UGent. FWO, the Research Foundation - Flanders has (co-)financed the underlying research of this publication conducted within the context of the Senior Research Project Confluvial: Towards an art-science-activist worlding activating awareness of (hydro)ecologies (PI, Christel Stalpaert, G093425N).

2 Natural Contract Lab prefers to use the combination of the French (Senne) and Dutch (Zenne) name to refer to the river, indicating that the river flows through the French-speaking and Flemish-speaking part of Belgium, crossing borders and language differences. Following the grammatical rule that one can never use an article before a person's name, I will refer to SZenne River instead of the more common formulation of *the* Senne.

3 I would like to sincerely thank Maria Lucia Cruz Correia, Vinny Jones, Lode Vranken, Jef Seghers and Brunilda Pali for their collaborative thinking in the Natural Contract Lab meetings, for reading this text and for providing valuable feedback.

4 Haraway introduces the term 'art-science-activist worldings' in different ways throughout her book. Most of the time, she doesn't use hyphens. Occasionally, she does (2016, 79). I am particularly drawn to her hyphenated use of the term. For it is precisely in the hyphens that non-conformist modes of knowledge production resurge.

References

Azoulay, Ariella Aïsha. 2019. *Potential History. Unlearning Imperialism.* London: Verso.

Ben Yakoub, Joachim. 2023. 'Breathing Out of the University, and into Many Schools: Storying Abolition as a Way of Activation.' *Performance Research* 8 (28), 109–19.

Common Dreams: Flotation School. 2017. "'Flotation School' on Survival, Time and Climate Change."

Facebook, 30 September. www.facebook.com/events/312879792513380/312879805846712
Correia, Maria Lucia Cruz. 2025. Email to the author, 25 March.
Correia, Maria Lucia Cruz. n.d. 'Common Dreams: Flotation School.' http://mluciacruz-correia.com/works/common-dreams-flotation, accessed 3 January 2025.
Ferdinand, Malcom. 2022. *A Decolonial Ecology. Thinking from the Caribbean World.* Hoboken, NJ: John Wiley & Sons.
Ferdinand, Malcom. 2024. *S'aimer la terre. Défaire l'habiter colonial.* Paris: Éditions du Seuil.
Ghosh, Amitav. 2021. *The Nutmeg's Curse: Parables for a Planet in Crisis.* Chicago, IL: University of Chicago Press.
Haraway, Donna J. 2016. *Staying with the Trouble: Making Kin in the Chthulucene.* Durham, NC: Duke University Press.
Latour, Bruno. 2005. *Reassembling the Social: An Introduction to Actor-Network-Theory.* Oxford: Oxford University Press.
Piña, Amanda. 2019. The School of the Jaguar: Rehearsing an ecology of knowledge. In *Endangered Human Movements.* Vol. 3: The School of the Jaguar, edited by Amanda Piña, 47–85. Nadaproductions.
Raunig, Gerald. 2013. *Factories of Knowledge: Industries of Creativity.* Translated by Aileen Derieg. Cambridge: MIT Press.
Stalpaert, Christel. 2020. 'De slagkracht van de verbindingsstreepjes.' *Collateral: Online Journal for Cross-cultural Close Reading.* June, Cluster 25. Thematic Issue on *Traag Geweld: kan Kunst het Klimaat redden?*, Stef Craps and Mahlu Mertens, eds. https://collateral-journal.com/index.php?collision=traaggeweld, accessed 3 January 2025.
Stalpaert, Christel. 2023. 'Performing Arts Activating Climate Change Awareness: Hyphenated Thinking in *Common Dreams — Flotation School.*' *TDR – The Drama Review: The Journal of Performance Studies* 1 (67), 74–81.
Stalpaert, Christel. 2025. 'Hyphenated Thinking in Performance Processes.' In Laura Karreman and Liesbeth Groot Nibbelink, eds, *Performance Research Methods.* Cambridge: Open Book Publishers, 409–434.

19

BURNING MAN'S CLIMATE

Eco-Utopic Confluences and Contradictions

Sarah Ann Standing

Burning Man—the yearly neo-pagan festival in the Nevada desert—embodies pressures of global climate change in microform: a desert scape with howling winds, extreme temperatures and water-shortage concerns. All attendees must bring their own drinking water for the week—about a gallon and a half per person daily. Unexpectedly, in 2023, there were monsoon rains. At *Burning Man*, towering tech-infused sculptures and a long relationship with Google founders[1] and employees imply an uneasy alliance with technology companies. Yet, with all its imperfections, *Burning Man*, now in its 39th year (as of 2025), has a following. 80,000 participants brave extreme environmental challenges to experience a complex, not wholly unmitigatedly, utopic gathering. This chapter examines *Burning Man* from an eco-utopic standpoint, attempting to unpack the complexity of the utopic festival as a different, more nuanced, impure notion of ecotopia, using an eco-performance lens and examining protest as a means of opening a gateway into thinking through *joy* as a valuable aspect of the aesthetics and politics of climate performance and climate response. Joy at *Burning Man* demonstrates a vital part of empowered activist climate response, as many people feel overwhelmed and in despair when looking at the science of climate change. A functional, appropriately joyful performance model that integrates the eco-utopic can help people re-invigorate and remain engaged in the face of despair. *Burning Man* is a highly imperfect model for this joy, but protest performs a counter to inaction, nonetheless. I am trained as a theatre historian, and here, I draw on various primary and secondary sources to situate *Burning Man* in the context of the eco-utopic.[2] This chapter reviews the ecologic history and tenets of the festival, investigates the interface with art and technology and, finally, addresses the question as to whether *Burning Man* and its utopic jubilation are,

DOI: 10.4324/9781003584445-25

indeed, eco-protest and how these aspects might contribute to protest studies and eco-performance studies.

Ecologic Zeitgeist of the Time

Burning Man began in San Francisco in 1986 on Baker Beach, which sits on the western edge of San Francisco—a city surrounded on three sides by water. Most of the city resides on the Bay side and inside the Golden Gate—the narrow deep-trench entrance to the San Francisco Bay. On the wilder, western edge outside the Golden Gate, Baker Beach is the liminal land between the city to the east and the Pacific Ocean stretching to the west with only a few rock outcroppings until Hawai'i—2,388 miles away. Baker Beach is a concealed cove of fine sand and pounding waves. Surfers need to wear wetsuits because of the frigid water, and they must remain vigilant for jagged rocks and rip currents. It is a place of bonfires, and in 1986, clothing was optional. This is a fitting place for the birth of Burning Man—a place of nature and culture, of creating community, of transgression and ritual.

San Francisco has long been a city with a free-spirited counterculture and was also an early pioneer in many aspects of the United States' mainstream environmental movement. These seem relatively tame by today's standards, yet in the 1960s and 1970s, they were harbingers of eco-philosophy and practice to come. They are essential indicators of the zeitgeist that gave rise to *Burning Man*. First, addressing post-consumer waste: as early as the 1970s, San Francisco developed ten voluntary community recycling stations,[3] and second, a growing awareness of vegetarian food and organic farming.[4] Third, Stewart Brand's *Whole Earth Catalog* was published by the Portola Institute in Menlo Park (in the San Francisco Bay Area), beginning in 1968. This was a tumultuous year in the United States with an escalation of the Vietnam War and the assassinations of both civil rights leader Martin Luther King, Jr., and leading Democratic presidential candidate Robert F. Kennedy. It was also the year following San Francisco's renowned 'Summer of Love'. According to Anna Wiener,

> At the height of the civil-rights movement and the war in Vietnam, the *Whole Earth Catalog* offered a vision for a new social order—one that eschewed institutions in favor of individual empowerment, achieved through the acquisition of skills and tools.
>
> *(2018, 12)*

This focus on individual empowerment dovetails with the ethos of the back-to-the-land movements of the 1960s and 1970s (intersecting with both the aforementioned sustainability and organic farming movements) and engendered a *Burning Man* project focussed on individual empowerment, defiantly distrustful of authority and decidedly non-materialistic. Additionally—and this is crucial—the 1970s earmarked the emergence of personal computers developed in the Bay Area of San Francisco, out of this very same ethos: to empower individuals in their work and creative visions. Although much of the development of computers in large companies in the United States was advanced by women and often women of color, early individual computers often came in kits designed for hobbyists. These kits were developed by and for local innovation hubs and mainly appealed to individuals who were like and like-minded: white, male, and tech-oriented.

Brief History of *Burning Man*

In the early 1980s, sculptor Mary Grauberger enjoyed a tradition of burning artworks on Baker Beach to celebrate the summer solstice. When she stopped, in 1986 Larry Harvey and Jerry James took over the tradition and started *Burning Man*, where 20 people gathered, built a bonfire, celebrated and ultimately burned a wooden man in effigy. After several years of this, the threat of arrest prevented the burning of the eight-foot-high sculpture, and the festival looked for a new venue. In 1990, *Burning Man* moved to Black Rock City, Nevada. The extra space and relative lack of civic oversight helped the festival grow so that by 1997, there were 10,000 attendees. Today, the festival hosts around 80,000 participants annually.

Black Rock City, Nevada, is a dry, former lakebed in the Black Rock Desert in Northwestern Nevada—playfully named 'La Playa' (The Beach) by *Burning Man* constituents. At around 400 square miles, it is one of the largest flat areas in the northern hemisphere and is full of extremely fine, alkaline dust. This dust is a pervasive presence, an aesthetic and intrinsic part of the tactile experience of the festival. The dust is so invasive that rental car companies in the Reno, Nevada, area have a specific clause in their contracts prohibiting taking cars to *Burning Man*, with cleaning fines of up to thousands of dollars for infractions. In 2023, amidst an unpredicted deluge, the dry lakebed reverted to a lake with, as some said, the consistency of peanut butter. These elements combine to produce an uber site-specificity, wherein the site tells the festival's story.

Utopia

Not only is there a focus on self-reliance at *Burning Man*, but there is also a deliberate emphasis on creating a utopic community. In 2004, co-founder Larry Harvey formalised ten 'principles' for creating this community. The ten principles are *Radical Inclusion*, *Gifting*, *Decommodification*, *Radical Self-reliance*, *Radical Self-expression*, *Communal Effort*, *Civic Responsibility*, *Leaving No Trace*, *Participation* and *Immediacy* (Burning Man 2025). Like the *Whole Earth*

Catalogue, these principles reflect a utopic vision predicated on self-reliance. However, the principles also build community by deliberately including strangers, establishing the principle of gift-giving without expectation of return and committing to non-commerciality. Still, the principles are not ordained in a hard and fast manner, nor are they prescriptive. According to the *Burning Man* website, these principles 'were crafted not as a dictate of how people should be and act, but as a reflection of the community's ethos and culture as it had organically developed since the event's inception' (Burning Man 2025). Thus, it is a utopia based on self-compliance, which I suggest can be another form of the historically important (especially in the United States) concept of self-reliance. Further, as Jill Dolan writes:

> Thinking of utopia as processual, as an index to the possible, to the "what if," rather than a more restrictive, finite image of the "what should be," allows performance a hopeful cast, one that can experiment with the possibilities of the future in ways that shine back usefully on a present that's always, itself, in process.
>
> *(2005, 13)*

Thus, the descriptive rather than proscriptive nature of the ten principles, continual evolution through the performance of the festival and imaginative future-creation through current-creation all cast *Burning Man* as the utopic, or utopic-aspirational—which applies to all utopias, as they are by definition impossible to achieve or sustain.

Eco-Utopia

The beginning of Burning Man on Baker Beach in 1986 was significant in terms of place and zeitgeist (as discussed previously) and time of year, originating as a Summer Solstice ritual. This veneration of the natural world through its initial connection to Paganism remains part of its stated ethos. One doesn't have to look far for intriguing Pagan antecedents of *Burning Man*. Burning a wicker man, for example, was cited by Julius Caesar witnessing Druid festivals 2,000 years ago, although corroborating evidence for this is scarce (Gilmore 2010, 32). These festivals continued for centuries in various forms in much of Celtic Europe. *Burning Man* is perhaps most directly linked to Beltane—a festival celebrated on 1 May—which prominently features fire and often the burning of an effigy. The connection between *Burning Man* and ritual has been written about extensively (see, for example, St John 2019; Gilmore 2010; Bowditch 2010). One of the ten principles of *Burning Man*, *Immediacy*, reads in part, 'We seek to overcome barriers that stand between us and a recognition of our inner selves, the reality of those around us, participation in society, and *contact with a natural world exceeding human powers*' (Burning Man 2025, emphasis added). Thus, a clear historical and contemporary connection exists between a spiritual experience with the natural world and an imported paganism.

These principles, in particular, *Immediacy*, *Radical Inclusion*, *Gifting*, *Decommodification*, *Radical Self-reliance*, *Communal Effort* and *Civic Responsibility*, speak to an eco-utopic vision. They refute late capitalism through decommodification, communal effort and generalised reciprocity. There is no money on La Playa. Everything (food, drinks, drugs) is exchanged through barter, but most things are free. The ten principles offer some functional eco-utopic modelling for moving forward together amid environmental crises and extreme geo-climate events.

Further, *Burning Man* can be considered an environmental festival as it involves ecological aspects, self-reliance, pagan beliefs in the form of respect for the natural world 'exceeding human powers', and the principle of *Leaving no trace*. *Leaving no trace* is described as follows:

> Our community respects the environment. We are committed to leaving no physical trace of our activities wherever we gather. We clean up after ourselves and endeavor, whenever possible, to leave such places in a better state than when we found them.
>
> *(Burning Man 2025)*

This is the stated goal, and people spend the week after the festival removing all trash and forgotten objects and sweeping La Playa. Eco-utopias are built on the foundational, although incomplete, principle of sustainability. This is one elementary feature that distinguishes eco-utopia from the original concept of utopia.

Art and Technology

The eco-utopia of *Burning Man* can be understood further through the principle of *Radical Self-expression*, which states that

> Radical self-expression arises from the unique gifts of the individual. No one other than the individual or a collaborating group can determine its content. It is offered as a gift to others. In this spirit, the giver should respect the rights and liberties of the recipient.

Creativity is seen as a directive at *Burning Man*. This addition, not only to the idea of utopia but also to the idea of eco-utopia, invites in the undetermined and as yet unknown, the 'processual' as Dolan would say (2005). It grounds the eco-utopia in the healthy expression of the self within the context of the community's health.

In addition to the ritualistic aspect, community and remote location, several other elements draw people to *Burning Man*. The scenography is undeniably evocative: the desolate landscape, light emanating from either participants or stars, the dust, the scale and the climax of combustion (the burning man) after a sweltering week under the sun. The festival is a celebration of outrageous, impermanent creativity in a 'natural' setting—and here, creators take ephemerality into their own hands and burn the man. Because these artworks are designed for the site, and the site is so particular (and the effects of the site so pronounced), we might say these artworks are 'co-produced nature-cultures' (Giannachi and Stewart 2005, 19). The artworks are an intrinsic part of *Burning Man*, as stated in the principle of *Radical Self-expression*, which 'arises from the unique gifts of the individual'. The artworks are usually large, spectacular and experiential. They often (although not always) involve automation, lighting and other theatre and computer technologies. They are almost universally interactive, and the ethos of gift-giving extends to the artworks themselves, so everyone is encouraged to gain something tangible, experiential or both from each artwork.

Every year, there is an evocative theme for the festival, which the artworks may choose to address. They are all to some degree 'positive' and utopian rather than dystopian, for example, *Fertility* (2012), *The Vault of Heaven* (2004) and *The Floating World* (2002). These sculptures are often movable and include art cars (motorised vehicles transformed into artworks and designed to be driven around), bicycle-powered vehicles and wind-powered kinetic sculptures. The area is so vast that it is helpful to have vehicles traversing the flat terrain. This also adds to the evocative dynamism of the large-scale artworks immersed in the desert scape. The interest from Google funnelled a lot of energy (and money) towards technology-infused art. In 2023, there was a 'Living Knitwork Pavilion' at *Burning Man*, sponsored by the Massachusetts Institute of Technology—where machine-knit fabric included 'photochromic, luminous, and conductive yarns'—and culminated in a lightshow (MIT Media Lab 2023a, 2023b). *Burning Man* has grown over the past 39 years to have its own 'Art Director', competitions for funding artworks, satellite experiences and live streaming of the event. This attests to the importance of art to the festival, as well as a level of sophistication and spectacle, not to mention commercialisation, which was not as prevalent in the early years.

Critique of the Festival

The festival presents itself as an anti-capitalistic communal utopia as systems of barter abound and a creed of self-reliance in service of building community exists in the liminal space. Everyone is a creator and performer. Yet, it is a predominantly white crowd, and it can be challenging to get a ticket. In later years, wealthy San Francisco tech executives fly in for a few days at a time. But most significantly, for our conversation here, *Burning Man* still insists on its mantra *Leave No Trace* and demands an environmental clean-up after the festival concludes. Is it possible for 80,000 people to leave no trace after living in a fragile desert environment for over a week? The *Leave No Trace* ethos, valuable though it is—and *Burning Man* is committed to returning La Playa each year to its former 'pristine' condition—refers to localised, visible pollution. Air and auto travel to reach the remote location are not factored in.

In response to the COVID pandemic, *Burning Man* was cancelled in 2020 and 2021. As a result, several other iterations of *Burning Man* sprang up. In 2023, 50 sculptures from *Burning Man* were imported for a special show at Resortworld on the Las Vegas 'Strip'. This is the boulevard that runs through the city of Las Vegas, featuring hotels, restaurants and famously themed gambling casinos. The show at Resortworld was entitled 'Transfix' and was dubbed 'The world's largest touring immersive-art experience' (Cascone 2023). The show was, by necessity, smaller than the experience on La Playa. Additionally, this 'immersion' refers only to an experience with technology but not an immersion in what we might call the nature/culture interface. Additionally and crucially, the Las Vegas show is designed for spectators, whereas La Playa applies a 'no spectators' code. The virtual *Burning Man*, BCRvr, which has been many years in the making, advertises itself as 'work[ing] year round with a fierce volunteer community creating truly immersive and expansive experiences in the metaverse' (BCRvr 2025). This use of virtual reality raises many considerations regarding the live vs. the mediated—a deep conversation in theatre studies. As stated previously, Google and *Burning Man* have a long relationship, and in 2023 collaborated on 'the process of selecting, creating and installing six significant artworks at the Google Visitor Experience in Mountain View, California' (Burning Man 2025). All of these facts flex the boundaries of the principle of *Decommodification* to the breaking point.

Because of the torrential rains in 2023, many news outlets dubbed *Burning Man* 'Drowning Man' or 'Mud-fest Man', which, paradoxically, put it on the front of many media organisations worldwide. I venture to say that it has never had such mainstream publicity. Arwa Mahdawi, writing for *The Guardian*, published an article provocatively titled 'Why All The Burning Man Schadenfreude? Where Do I Start…' (2023). Although there are many plausible reasons for the deep divide between people who find the festival an essential part of their lives and those who see it as annoying or even pretentious, Mahdawi takes on the resentment against tech leadership flying into Black Rock City in private planes. The tech leadership violates core principles of inclusion, community and gifting by glamping, bringing their high-end chefs, and not participating equally with others. Mahdawi reminds us that the early participants in *Burning Man* were members of the counterculture. Adding on to Mahdawi's perspective, it is important to remember that the founders of much of the technology industry were also counterculture outsiders—imagining computers for individual home use and seeing art and technology as intrinsically interconnected. This is before being co-opted into a driving force of neo-liberalism and surveillance. It is this outsider spirit that *Burning Man* continues to embrace and advocate. In many ways, it is a community of outsiders conjoined by the vital force of joy. *Burning Man* presents an eco-utopia conjoined with a joyful and exuberant use of technology, refuting the false divide between technology and sustainability often mandated by eco-utopists.

Perhaps festival goers feel true schadenfreude because the 'Transfix' show closed months ahead of schedule, and the organisers eventually filed for Chapter 7 bankruptcy (a legal proceeding in the United States and a means to discharge a company's debt when they are unable to pay their bills). *Burning Man* is now fully entrenched in popular culture, and further evidence of the irksomeness of festivalgoers is detailed in *The New York Times* in, of all places, the 'Modern Love' section. As Twining-Ward writes, 'The promise of inclusion, sustainability and self-expression is alluring. Social media has manufactured a generation of alienated, wasteful and insecure young singles desperate for tangible meaning'. Yet, she says, the self-absorption and the self-righteous blindness to privilege of many 'burners' is overwhelming (2024). This is another common problem in utopic communities where people tend to 'drink the Kool-Aid'.

Protest

I return to the last of the ten principles, *Immediacy*, which, as stated previously, culminates with 'No idea can substitute for this experience'. The idea of 'overcome[ing] barriers that stand between us and a recognition of our inner selves' seems to echo Norwegian eco-philosopher Arne Naess's Deep Ecology with its focus on transforming the self as an ecological act. It is this focus on embodiment that I am driving towards in this chapter. In 2023, I participated in the 'La Selva': International Ecological Arts and Eco-Somatic Residency in Rome, Italy. There, an emphasis on physical knowledge and physical interaction as a source of knowledge made me think about the experiential aspects of *Burning Man*. The art does not mimetically represent the Anthropocene but an experience of interacting with and celebrating what I call the 'Mad Maxification' of the dust-saturated environment and inescapably interacting with the landscape in an embodied experience. The art cars often resemble vehicles in the Mad Max film franchise, and the setting is eerily reminiscent. The intense heat and dust of the desert (most years), the music, the Playa and the food and drinks gifted are part of the experience. In 2023, sudden torrential downpours turned the Playa back into a lake or a mud bath. Vehicles were stuck—axle deep—and there was a shelter-in-place order for several days. One person died. This caused much notoriety (as previously mentioned) but also gave a diametrically opposed physical experience—but one no less all-consuming.

In answer to the question: is this an eco-protest? I answer yes, although protest of a different kind than we might usually think about. It is not the kind of direct action eco-protests practiced by, for example, Greenpeace, Earth First! and Extinction Rebellion, and as such it is perhaps less recognisable as eco-protest. It does not carry a focussed, overtly political agenda addressing a specific issue. *Burning Man* draws a larger proportion of white men who earn over $100,000 a year, than most protests. What places it in the protest category? The original meaning of the counterculture was counter to the military-industrial complex (although much of the technology was developed in conjunction with the military-industrial complex, as well as universities). Thus, although *Burning Man* flexes the definitions and stretches the boundaries of protest, even as it vexes many in the hypocrisy and pretension it embodies, I argue that it is a highly useful form of protest nonetheless because it reminds us of the value of jubilant engagement, embodied within an environmental and social context.

Pranks, humour and jubilant engagement in protest have a long history. The San Francisco Mime Troup, founded by R.G. Davis, used jubilation, humour and, significantly, free shows to expound class equality, environmental justice and counterculture. The Diggers (also from San Francisco) used gift-giving in the form of free bread, a Free City Bank and the *Free City News*. Both the San Francisco Mime Troupe and The Diggers were political theatre groups that performed playful, exuberant work with a serious message and were clear forerunners to *Burning Man*. James Thompson (2009) writes that beauty and joy are legitimate and necessary aspects of political theatre. *Burning Man*'s Dionysian jubilation, in all its paradoxical hypocrisy, still generates a vital level of transgressive joy and excitement—making participation an act of devotion and defiance—amidst extreme geo-climatic conditions. This vital joy is essential to combat the overwhelming despair many feel in the face of climate destabilisation. Peta Tait poignantly writes: 'The redemptive possibility of artistic collaboration and creativity counterbalance how intersecting fiscal, political and sensory economies induce individual moods of despair for the future' (2021, 4).

Alice Walker (1992) writes that resistance is the secret of joy. This chapter asks if joy is not also a secret to resistance? The *Burning Man* movement, though it seems to be apolitical, can be said to embody eco-resistance and a kind of protest against an isolated life, a non-creative life, a non-communal life, a life of known experiences, but also, most importantly, a life overwhelmed by 'capitalist consumption'. This feeling of being overwhelmed falls upon many who engage with the reality of climate destabilisation, the oil oligarchy and the extreme violence undertaken to protect the fossil fuel industry as a source of wealth. Despite its inevitable contradictions and problems, the vitality of

Burning Man—this transgressive, irrepressible joy—also warrants a place in our conversation about climate resistance and how we move forward together.

The utopic is often depicted as an unobtainable ideal that exists no-where. The very force of La Playa is an element to contend with (we haven't even talked about sanitation yet), but I argue that a reclamation of *place* in what Chaudhuri terms a 'materialist-ecological theatre' (1994, 24) as an aspect of the utopic is a contribution *Burning Man* can make to climate performance studies. The exuberant joy of *Burning Man* conjoins with a fully embodied connection to place—no longer embracing the Platonic ideal as summarised in 1516 by Sir Thomas Moore, of utopia as no-where and no-thing but abstract thought (2012 [1516]). Instead, *Burning Man* is a worthwhile, joyous, jubilant embrace of the earthly utopic.

One could say that this earthly utopic perspective dovetails with many Indigenous or eco-feminist perspectives, and yet *Burning Man*, emanating out of a white, masculinist, technology-infused zeitgeist, becomes a study of the contradictions in utopic eco-protest. These contradictions should not be easily dismissed, however, as they mirror confluences of current global power, as witnessed in the US presidential election of 2024.

Notes

1 In 1999, Google founders Larry Page and Sergey Brin went to *Burning Man* with a number of the company's employees. They created a *Burning Man* logo for Google's homepage to let users know that leadership would be away from the company. They became longtime attendees.

2 I am grateful to Carl Lavery for his insightful comments on an earlier draft of this essay.

3 In 1996, there was city-wide composting; in 2003 the Zero Waste Ordinance, and in 2009, the Mandatory Recycling and Composting Ordinance was enacted. Today, San Francisco has 80 per cent waste diversion, the highest percentage in the US, and higher than most cities globally.

4 Frances Moore Lappé's 1971 book *Diet for a Small Planet* took the US by storm. In 1969, San Francisco opened Real Foods Grocery Store, which featured organic foods directly from small farmer produce; in 1971, Alice Waters opened Chez Panisse in Berkeley. In 1979, Greens was founded by the Zen Center and drew produce from their Green Gulch Farm.

References

BCRvr. 2025. 'The Virtual Burning Man Experience.' https://brcvr.org/about-brcvr, accessed 15 March 2025.

Bowditch, Rachel. 2010. *On the Edge of Utopia: Performance and Ritual at Burning Man*. New York: Seagull Books.

Brand, Stewart. 1968. *Whole Earth Catalog*. Menlo Park, CA: Portola Institute.

Burning Man. 2025. 'The 10 Principles of Burning Man'. https://burningman.org/about/10-principles, accessed 17 April 2025.

Burning Man. 2025. 'Programs: Burning Man Arts, The Google Visitor Experience Project. https://burningman.org/programs/burning-man-arts/civic-arts-program/gradient-canopy/faq/, accessed 19 September 2025.

Cascone, Sarah. 2023. 'The Largest Touring Immersive Art Experience Is Bringing 50 Burning Man Style Sculptures to Las Vegas Before It Hits the Road.' *Artnet*, 1 March. https://news.artnet.com/art-world/transfix-immersive-art-experience-las-vegas-2262410, accessed 15 March 2025.

Chaudhuri, Una. 1994. '"There Must Be a Lot of Fish in That Lake:" Toward an Ecological Theater.' *Theater* 25 (1), 23–31.

Dolan, Jill. 2005. *Utopia in Performance: Finding Hope at the Theatre.* Ann Arbor: University of Michigan Press.

Giannachi, Gabriella and Nigel Stewart, eds. 2005. *Performing Nature: Explorations in Ecology and the Arts.* Bern: Peter Lang.

Gilmore, Lee. 2010. *Theater in a Crowded Fire: Ritual and Spirituality at Burning Man.* Berkeley: University of California Press.

Lappé, Frances Moore. 1971. *Diet for a Small Planet.* New York: Ballantine Books.

Mahdawi, Arwa. 2023. 'Why All the Burning Man Schadenfreude? Where Do I Start….' *The Guardian*, September 5. https://www.theguardian.com/culture/commentisfree/2023/sep/05/why-all-the-burning-man-schadenfreude-where-do-i-start-, accessed 17 April 2025.

MIT Media Lab. 2023a. 'Living Knitwork Pavilion @ Burning Man 2023.' https://www.media.mit.edu/events/living-knitwork-pavilion-in-burning-man-2023, accessed 17 April 2025.

MIT Media Lab. 2023b. 'From MIT to Burning Man: The Living Knitwork Pavilion.' 13 October. https://www.media.mit.edu/articles/from-mit-to-burning-man-the-living-knitwork-pavilion, accessed 17 April 2025.

Moore, Sir Thomas. 2012 [1516]. *Utopia: A Visionary Exploration of Equality, Justice, and the Ideal Society.* London: Penguin Classics.

St John, Graham. 2019. 'At Home in the Big Empty: Burning Man and the Playa Sublime.' *Journal for the Study of Religion, Nature & Culture* 13 (3), 286–313.

Tait, Peta. 2021. *Theory for Theatre Studies: Emotion.* London: Methuen.

Thompson, James. 2009. *Performance Affects: Applied Theatre and the End of Effect.* Basingstoke: Palgrave Macmillan.

Twining-Ward, Cate. 2024. 'Men, Please Stop Talking about Burning Man: Am I the Only Woman Meeting Burning Mansplainers on Dates?' *The New York Times*, 8 November. https://www.nytimes.com/2024/11/08/style/modern-love-burning-man-mansplainers.html, accessed 28 April 2025.

Walker, Alice. 1992. *Possessing the Secret of Joy.* New York: Harcourt Brace Jovanovich.

Wiener, Anna. 2018. 'The Complicated Legacy of Stewart Brand's "Whole Earth Catalog".' *The New Yorker*, 16 November. https://www.newyorker.com/news/letter-from-silicon-valley/the-complicated-legacy-of-stewart-brands-whole-earth-catalog, accessed 17 April 2025.

20

TACUABÉ'S BOW

Surviving through Absence

Maria Litvan

The Tacuabé bow is an ancient Charrúa musical instrument. The Charrúas were the dominant Indigenous peoples of the land now known as Uruguay. They were known for their courage and great ability in battle, so much so that they were enlisted to fight against Brazilian and Argentinian domination and were instrumental in helping Uruguay gain its independence. In 1831, in a seeming act of appreciation, Fructuoso Rivera, the newly established nation-state's first president, asked every Charrúa cacique (tribal chief) to gather their people (women and children included) to convene at the stream of Salsipuedes, promising them land and cattle (Houot 2013, 14). Rivera, a military man, had personal relationships with the Charrúa caciques, having fought with them against foreign domination. On 11 April of that same year, most Charrúas responded to the president's invitation. While a minority chose not to go, suspicious of the president's intentions, those who attended were received with gifts and alcohol along with friendly gestures (Houot 2013, 14–15). But when the Charrúas let down their guard, and the men became inebriated, the Uruguayan soldiers fired on them. Salsipuedes, the name of the stream and its surrounding area, has become a synecdoche for the genocidal decimation of the Charrúa population. A few Charrúas were able to escape, and the surviving Charrúas—mostly women and children—were taken by foot to Montevideo (a walk of more than 180 miles), where they were sold as slaves to wealthy families. Children were separated from their mothers. The intention was to eradicate the Charrúa language and culture. By these genocidal means, Uruguay established itself as a country with no 'Indians'—an exception in the Americas.

European invaders not only attempted to eradicate the Indigenous peoples and their cultures, but they also introduced a new economy that would forever

DOI: 10.4324/9781003584445-26

change the land and its biodiversity. They favoured a system of monoculture that continues today to exhaust the soil, leaving it vulnerable to infestations of New World Screwworm flies and other pests. Just as European colonists justified their taking of the land by dehumanising Indigenous peoples and claiming that the land was empty (Terra Nullius), Uruguay's grasslands—characteristic of the Pampa biome, shared with Argentina and southern Brazil—are overtaken by monocultures, mostly of industrial eucalyptus and pine trees, as well as soybean crops. Despite their rich biodiversity, because of their lack of trees, grasslands are perceived as being empty (Lagos 2022). While tree planting is promoted worldwide to mitigate climate change, the planting of trees in grassland ecosystems severely damages their biodiversity and negatively impacts waterways (Selva Ortiz and Achkar 2023). The effects of this forceful occupation of a land and its peoples have trickled into what now appears as an environmental disaster. In May 2023, Uruguay suffered an unprecedented water crisis that kept almost half of the country's population without access to drinking water for over 60 days (World Rainforest Movement 2023).

I argue that through 'disappearance' (by making themselves invisible, which is very different from eradication), the ancestral Charrúas safeguarded the possibility for new futures. Absence (understood not as a lack, but as full and creative) has played a more active role than presence (as in evidenced present, material occupation of spacetime) in the Charrúas' processes of preserving their cultural heritage. This account of heritage through Tacuabé's bow is in contrast to a colonial monoculture. This chapter speaks for a diverse cultural ecology and proposes performance, in particular Tacuabé's bow, as a practice to access other epistemologies and ways of being in the world, other ecologies.

Charrúas Reclaim Their Ancestry

In the late 1980s to early 1990s, following the end of the 1973–85 military dictatorship, the descendants of the Charrúas slowly started to reclaim their ancestry. For the first time, associations were established to fight for recognition of lineage and to contribute to the recovery and preservation of Indigenous culture and identity. However, whereas there may be recognition of the right to self-identification, the state does not yet recognise those who self-identify as Charrúa as the originary people of the land. Anthropologists Renzo Pi Hugarte (1934–2012) and Daniel Vidart (1920–2019), considered the 'founding fathers' of Uruguayan anthropology, continued to negate the possibility of the prevalence of the Charrúa lineage.[1] I was born in Uruguay, but due to the military dictatorship, my parents and I went into exile when I was only two years old, and I grew up in Barcelona. I have always been very curious about the Charrúa people but, since I was a little girl, every time I inquired about them, I came up against the 'official' story that they were now extinct, and there was hardly any information about them. As anthropologist Andrea Olivera (2016, 47) states, 'when we look for the Charrúas, we encounter a street without an exit. The Charrúa nation seems to be an invisible nation, or at least, a nation present for its absence'.[2]

The Charrúas, along with other tribes that resisted assimilation, were seen and described as violent and barbaric savages. Ironically, they were depicted as robbers, bandits and rapists by those who did the ultimate robbing and raping of a land and its peoples. Such questionable depictions have nevertheless created and perpetuated the discourse that the Charrúas were a violent tribe, so primitive that they were incapable of producing a single architectural structure, and consequently left no valuable legacy for the modern world. 'They were not like the Aztecs or the Maya or the Inka, the Charrúas were a crazy violent bunch who left nothing' is what I heard again and again as I was growing up, and my curiosity, and maybe something more than curiosity, insisted that I keep asking about them. But the message was always the same: there was nothing I could know or learn from the Charrúas. They were gone, leaving behind no more than a few 'boleadoras' [bolas, a hunting weapon] and 'puntas de flecha' [arrowheads]. But, despite the erasure performed against them, in December 2021, I met Ciro: a happy, charismatic and full-of-life Charrúa. And then, I met several other Charrúas, and I learned that not everything was lost. Then I discovered I knew (Uruguayan) Charrúas in New York! I was and remain stupefied by the dimension of the lie that still pervades the common Uruguayan consciousness. The Charrúas continue to be very much present and alive, but they had to make themselves invisible to be able to survive.

I argue against the widespread acceptance that the Charrúas were a group of violent barbaric warriors, that they, even if fierce warriors, were a meditative tribe, not interested in conquering external space. Furthermore, once contact with the Europeans took place, the Charrúas were heavily persecuted. They had to be ready to escape at any moment, rendering them even more nomadic and limiting their engagement in producing and possessing material objects (Pi Hugarte 2018, 44; Sanz 1955, 12). This chapter claims that the Charrúas transmitted their culture through meditative engagement and merging with their environment, rather than occupying it with architectural structures and other artefacts.

El arco de Tacuabé (Tacuabé's bow)

Since 2022, I have been working with Ciro, who is the cacique of the clan Choñik and now also president of CONACHA (Consejo de la Nación Charrúa/Council of the Charrúa Nation), to learn the methods that he and the Choñik use to reemerge their ancestors' culture and to (re)establish their Charrúa identity through performance and ritual practices. Ciro is a musician, and one of his main projects is the dual recovery of the ancestral musical instruments and the Charrúa language (see Figure 20.1).[3] There is one instrument, *el arco de Tacuabé* [Tacuabé's bow], that for me epitomises what I am coming to learn about the Charrúas but is missing from official accounts. The ancestral Charrúas resisted acculturation, and they kept their culture inaccessible to those who came with dreams of conquest. Thus, through Western epistemic lenses, they 'left nothing'. Yet, there is a rich heritage accessible to those who know how to listen differently. Engaging deeply with the musical bow might prove a closer way to approach the ancestral Charrúas and their ways of being in the world.

Tacuabé's bow is a musical instrument that, instead of projecting music outwardly to delight its listeners, is played to induce in its player a highly meditative state. The instrument takes its name from Tacuabé, one of the so-called 'last four Charrúas' taken to Paris in 1833 by François de Curel. De Curel, a former French sergeant, had asked Fructuoso Rivera, Uruguay's newly elected first president, if he could take a few 'Indians' with him to France for anthropological purposes and to exhibit the remains of a race now in extinction (Ayestarán 1953, 21). The proposal to send four of the strongest survivors of the Salsipuedes's genocide across the ocean must have seemed ideal to Rivera, given the threat the Charrúas posed to the newly established Uruguay. Rivera agreed. On 7 May 1833, after more than two months of sailing, the French brig Phaéton arrived in France. On board, there were 14 French crew members and 13 passengers, all of them also French except for four Charrúas: the cacique Vaimacá Perú; the healer Senaqué; the young Laureano Tacuabé, a horse tamer who was considered to be very dangerous and thus shackled throughout the journey (Houot 2013, 55); and one Charrúa woman, Micaela Guyunusa, who was pregnant (Houot 2013, 55). The four travelled on a single passport, and because France had a law forbidding entry of anyone against their will, François de Curel carried an official document from the ministry stating that the four Indians were following him of their own will (Houot 2013, 11). Obviously, Tacuabé's shackles contradicted such a document, but the French did not seem to have cared. The Uruguayan state—turning a blind eye to all the remaining Charrúas—pronounced them to be the last four. This is how they were recorded in the official history of Uruguay as it sought to create the idea of erasure and extinction of the tribe. An idea that, as mentioned earlier, has had such a strong imprint in the collective consciousness that even those who began to finally emerge as Indigenous in the 1990s were at first calling themselves 'descendants' of the Charrúa people, since the Charrúas were all dead.

Upon arrival in France, de Curel exhibited the four Charrúas in a residency on the Allée d'Antin [Antin Alley] in Paris. While the general public had to pay a fee, he invited the members of the Academy of Natural Sciences to examine the Amerindians free of charge (Rivet 2002, 39). It only took a few months for Senaqué to die from what was said to be 'a wasting fever' caused primarily by 'hopelessness, boredom and, above all, nostalgia' (Rivet 2002, 44).[4] Vaimacá Perú also died shortly after. Some of the French scientists, appalled by the conditions in which the Charrúas were kept, ended up reporting de Curel to the police.

It seems that de Curel learned about the report early enough and was able to escape to Lyon with Tacuabé, Guyunusa and her newly born daughter. In Lyon, Guyunusa died of pulmonary consumption and we lose track of Tacuabé and the baby (Rivet 2002, 56). Anthropologist and historian Darío Arce Asenjo (2007) found the death report of a Caroline Tacouavé, whom Asenjo claims is Guyunusa's daughter. If so, the baby died only a year after her mother, also in Lyon.

According to the ethnomusicologist Lauro Ayestarán (1913–66), Uruguay's most celebrated musicologist, a sentence in Félix de Azara's renowned *Viajes por la América Meridional* [Travels through South America, originally published in 1809 in Paris] stating that the Charrúas 'do not have games, dances, songs, musical instruments, or leisurely conversations' propagated the idea that they were a people without music (de Azara 1943, 106; Ayestarán 1953, 5).[5] Ayestarán worked to dismantle the myth and pointed to an essay of 1833 by phrenologist Pierre-Marie-Alexandre Dumoutier in which the author offers a detailed account of the musical instrument that Tacuabé took with him to Paris.[6] Dumoutier not only describes Tacuabé's musical bow but also the manner in which it was played (Ayestarán 1953, 23). It is known that Guyunusa and other Charrúas also played the instrument, but because Tacuabé happened to take it to France, today in Uruguay the musical instrument is known as 'el arco de Tacuabé' [Tacuabé's bow]. The bow is made with a 'fairly straight branch' of an autochthonous tree and a single string that is tied to each end of the branch, bending it to form a bow (Dumoutier in Ayestarán 1953, 23). Dumoutier (in Ayestarán 1953, 23) refers to it as a kind of monochord violin, 'from which I've seen them make very sweet and quite harmonious sounds'.[7] As we see in the image below, the player takes the instrument with one hand on the side of the branch that serves as the instrument's handle and holds the other side of the branch in-between their teeth. With a smaller and completely straight stick (also from the same tree branch) that, according to Dumoutier, is wetted with saliva, the player produces vibrations by striking the string. The opening and closing of the player's mouth help 'modulate and vary the tone' (Dumoutier in Ayestarán 1953, 23).[8]

In Dumoutier's account, the string is made with hairs from a horse's mane (around 15–20 hairs would be used), but, prior to the arrival of the Europeans and their horses, the Charrúas used ñandú (or *Rhea americana*)'s guts. *Ñandú* is a Guaraní word, *vera* is in Charrúa and Uruguayans use the Guaraní word to refer to the animal. Today, a guitar string is used instead so that the instrument sounds a little louder; it projects more outwardly than the ancestral version. Yet, it continues to vibrate in the player's mouth and thus maintains the original inward-directed purpose of the instrument. In Uruguay, Tacuabé's bow has been used to teach music to the hearing-impaired.

The monochord musical bow is not unique to the Charrúa people. According to Ayestarán (1953), it was the first chordophone known to humans.

FIGURE 20.1 Ciro plays Tacuabé's bow during his visit in October 2023 to the Graduate Center, CUNY. Photograph by Matia Litvan.

He describes two types of musical bow: those that have a resonance box and those that resonate in the mouth. Tacuabé's bow belongs to this latter category. Ayestarán (1953, 30) argues that when the musical bow has no resonator, it is only audible to the player and it is used to 'talk to the spirits'.[9] The ancestral Tacuabé's bow, also according to Ayestarán, is only audible within 20 centimetres of its player. The Guaraní, another group of Indigenous people sharing land with the Charrúas, have a similar bow that they play with a stick that is rubbed against the string. Without placing it in their mouth, it is closer to a European violin. Ciro tells me that the form of musical bow that resonates in the mouth, like Tacuabé's bow, is associated with tribes of hunter-gatherers who are venerators of the moon, like the Charrúas. Ciro jokes that it is dangerous to play Tacuabé's bow because 'when you play and play the bow, you begin to dance with the bow. Then, mind is no more. Everything becomes body and consciousness'.[10]

Tacuabé's Bow is embedded in the Charrúa culture. It evokes the profound and symbiotic relationship between the Charrúas and their environment. The land for the Charrúas is not a resource to be exploited but a living entity and an aspect of their own being (the foundation of their identity, the keeper of memory and spiritual connection). The bow's fate, caught up in European colonialism, points to what was lost in the more ecological practices that preceded the arrival of the colonists. Tacuabé's bow was almost lost, like the land that suffers from climate change caused in part by large-scale farming and a system of monoculture that strips the land of its biodiversity. It is no coincidence that the reemergence of the bow and of Charrúa culture and identity in Uruguay coincides with ecological efforts to protect waterways and promote agroecological food production. Agroecology in Uruguay, with its focus on social justice, reestablishes the connection between land and its peoples. It seeks to build sustainability beginning in the agrarian and extending to society as a whole (Gazzano and Gómez Perazzoli 2017).

Transmission takes place through other, more subtle paths, than what the modern world wants us to believe. To survive and continue into posterity, rather than focussing on production and on imposing our knowledges and believe-systems onto others, and our environments, we might need instead to learn to listen more carefully—as Tacuabé's bow suggests by inducing an inward state of attention and receptivity. My provocation here is to leave us thinking about these other possibilities of being and relating with everything around us; to offer a halt to the modern obsession with presence and visibility. I believe that in activating these other ways of relating to the world lies a more honest, and possibly more effective, means to decolonise thought. Also, a more hopeful alternative way of living with each other. And maybe, like it was for the Charrúas, the only way to survive through what appears as an impending extinction.

Notes

1 Until recently, in Wikipedia, the Charrúas still appeared as 'ethnically extinct'. On 25 January 2023, the Wikipedia page was corrected to acknowledge that there were surviving Charrúas communities outside of Uruguay (in Argentina and Brazil), and that 'It is believed that there are approximately between 160,000 and 300,000 individuals in Uruguay, Argentina, and Brazil today who are descendants of surviving Charrúa' (Wikipedia 2024).

2 'Cuando buscamos a los charrúas nos enfrentamos a un callejón sin salida. La nación charrúa parece ser una nación invisible, o al menos presente por su ausencia'. [All translations from Spanish to English are the author's.]

3 Ciro was invited by the Martin E. Segal Theatre Center to perform the ritual-performance *Indio Espinosa: A Multitemporal Charrúa Legend* in the 2023 *Prelude Festival*. His visit coincided with a gathering among CUNY students, faculty, and staff organising to request an Indigenous Studies Program within New York's largest public university. We invited Ciro to perform at the gathering and he played Tacuabé's bow.

4 'Una "fiebre de consunción", causada principalmente por "la desesperanza, el tedio y, sobre todo, la nostalgia"'.

5 'No tienen juegos, bailes, cantares, ni instrumentos músicos, tertulias ni conversaciones ociosas'.
6 Ayestarán found Dumoutier's essay in French ethnologist M. Paul Rivet's *Les Derniers Charrúas* [The Last Charrúas] published in 1930.
7 'Del que les he visto arrancar sonidos muy dulces y bastante armoniosos'.
8 'Modular y variar el tono'.
9 'Conversar con los espíritus'.
10 'El arco cuando vos tocás y tocás empezás a danzar con el arco. Ya no es más mente, ya es todo cuerpo y conciencia, por eso digo que es peligroso'.

References

Arce Asenjo, Darío Arce. 2007. 'Nuevos datos sobre el destino de Tacuavé y la hija de Guyunusa.' In

Romero, Sonnia. (comp. Edit.) *Anuario de Antropología Social y Cultural en Uruguay*, 51–71. Editorial Nordan-Comunidad. FHCE, UDELAR.

Ayestarán, Lauro. 1953. *La Música en el Uruguay*. Volume 1. Montevideo: Servicio de diffusion Radio Eléctrica.

de Azara, Félix. 1943. *Descripción e Historia del Paraguay y del Rio de la Plata*, 1st Argentinian ed. Buenos Aires: Editorial Bajel.

Gazzano, Inés and Alberto Gómez Perazzoli. 2017. 'Agroecology in Uruguay.' *Agroecology and Sustainable Food Systems* 41 (3–4), 380–400. https://doi.org/10.1080/21683565.2017.1286533.

Houot, Annie. 2013. *El trágico fin de los indios Charrúas*. Montevideo: Ediciones Cruz del Sur.

Lagos, Leo. 2022. 'Pastizales un ecosistema olvidado a la hora de hablar de conservación.' *La diaria ciencia,* 14 June. https://ladiaria.com.uy/ciencia/articulo/2022/6/pastizales-un-ecosistema-olvidado-a-la-hora-de-hablar-de-conservacion/, accessed 29 April 2025.

Olivera, Andrea. 2016. *Devenir charrúa en Uruguay: una etnografía junto con colectivos urbanos*. Translated by Darío Arce. Montevideo: Lucida Ediciones.

Pi Hugarte, Renzo. 2018. *Los indios del Uruguay*, 3rd ed. Montevideo: Ediciones de la Banda Oriental.

Rivet, Paul. 2002. *Los últimos Charrúas*. Translated by Mónica Sans, 2nd ed. Montevideo: Ediciones de la Plaza.

Sanz, Victor. 1955. *La propiedad en el charrúa*. Montevideo: Hispania.

Selva Ortiz, María and Marcel Achkar. 2023. 'El impacto de la forestación en el déficit hídrico.' *Brecha*, 17 February. https://brecha.com.uy/el-impacto-de-la-forestacion-en-el-deficit-hidrico/, accessed 1 May 2025.

Wikipedia. 2024. 'Charrúa.' https://en.wikipedia.org/wiki/Charrúa, accessed 9 February 2024.

World Rainforest Movement. 2023. 'Uruguay: The Impact of Industrial Tree Plantations on the Water Crisis.' *WRM Bulletin* 267, 25 October. https://www.wrm.org.uy/bulletin-articles/uruguay-the-impact-of-industrial-tree-plantations-on-the-water-crisis, accessed 1 May 2025.

The Recalcitrant Politics of Climate Change

21

CLIMATE CONTROL

Gender, Nature and Stories of the Snow Queen

Clara Wilch

No region is more significant to the history of climate imaginaries or the immediate material effects of climate change than the Arctic. For decades, the Arctic has been warming and transforming at a rate up to four times faster than the rest of the globe, and for decades, it has been at the centre of international climate communication campaigns (Rantanen et al. 2022).[1] How the Arctic is understood and valued is significant to the climate futures we collectively sow.

I've spent time in the northernmost Canadian territory of Nunavut, speaking with performers about place, climate, politics and their work. Browsing a general store in the capital city of Iqaluit once, I found myself facing a shelf of merchandise featuring Elsa, the Disney animated character. Having travelled from Los Angeles (the heartland of Disney Studios) with an interest in the global influence of material and imaginary polar icescapes, I was amused (but not very surprised) to see this reminder of multi-directional influences, a California-based vision of the Far North in global circulation. It was high time, I realised, to grapple with the extraordinary aesthetic reach and narrative power of the Snow Queen.

Disney's *Frozen* franchise has a globally dispersed and pronounced influence on Arctic imaginaries. The fictionalised kingdom of Arendelle, in which the animated films are based, is modelled on Scandinavian (primarily Norwegian) sites and aesthetic traditions (ABC News 2019), and the plot centres on a dramatic journey further north. *Frozen*, the first movie in what will be (at least) a trilogy, was the most profitable animated film ever at the time of its release in 2013 (Konnikova 2024), and the films are only part of a transmedia project that exists across many dramatic forms. These include a touring Broadway theatrical production; theme park rides; and a proliferation of merchandise,

DOI: 10.4324/9781003584445-28

all of which influences play and everyday behaviours for generations of fans (see Hewes et al. 2016). This transmedia performance infrastructure contributes to *Frozen*'s commercial force, popular longevity and international reach, which is difficult to overstate (and almost certainly familiar to readers.) Because weather and climate are central to the plot of *Frozen*, its environmental associations are also inescapable, exceeding the ties between a factual and fictionalised Arctic. The connection is so evident that an envoy from the Obama-era US State Department approached a Disney executive with a proposal to use the film's characters to teach children about climate change. The 'perplexed' Disney representative purportedly responded 'no' because 'it's in our culture to tell stories that project optimism and have happy endings' (Cama 2017). Yet, regardless of corporate or creative intentions, *Frozen* has naturalised environmental logics for its massive audience. This chapter examines those environmental logics and how they intersect with our present socio-climatic conditions (Figure 21.1).

FIGURE 21.1 A young person dressed in an Elsa costume looks out over the snow to the horizon. Image courtesy of Dan Weiss.

Popular and academic critics often read *Frozen* in terms of gender and sexuality, focussing on the ways that the identity expression and journey of its protagonist, Elsa, diverges from the marriage plot of most of Disney's princess films. This chapter instead focusses on the thermal narrative that is intertwined with the figure of the Snow Queen across iterations of storytelling. I ask what the gendering of ice and cold signifies in the context of contemporary climate change, beginning with the logics of gender and of nature in the source fairy tale.

Frozen is an adaptation of Hans Christian Andersen's story 'The Snow Queen' (or 'Snedronningen' in the original Danish), first published in 1844. Despite great departures from this text, key elements of *Frozen*'s narrative, characters and aesthetics are rooted in Andersen's work. Andersen was a contemporary of the Brothers Grimm, but unlike those collectors of Germanic folktales, Andersen largely wrote original fictions in the narrative template of fairy tales (Holbek 2002, 149). Andersen's works have since become folkloric, serially adapted mainstays of children's literature, and the Snow Queen has become the archetype of an enduring understanding of feminine power and mystique.

Andersen's story begins with the shattering of a mirror that was created by the devil to amplify all that is bad and diminish all that is good. Now in countless splinters, the mirror scatters around the world and pierces unsuspecting people, distorting their perceptions. Two young neighbours and best friends, Kai and Gerda, are introduced to readers shortly before Kai is infected by the mirror, which pierces his eye and his heart. His heart begins to harden into ice, and he becomes critical and cruel to Gerda. Kai goes ice-skating in the town's square and has a chance encounter with the Snow Queen. Kai hooks himself to the Queen's sleigh for a joy ride; she offers him a fur for warmth and kisses that freeze him, and she takes him to her ice palace very far away.[2] They keep each other company, but he is largely lost, pondering puzzles he cannot resolve. Gerda goes on a long and daring adventure to recover Kai, encountering many aids and obstacles. When Gerda reaches Kai, her tears (and then his) wash away the mirror splinters, and she and her newly liberated friend return to their home. Now adults, they are in love but still bathed in a child-like innocence. Andersen's story closes on the two, sitting side by side, enjoying their grandmother's rendition of a psalm about Jesus and roses.

In a story that repeatedly connects cold with logic and warmth with emotion, the Snow Queen acts like an equation between environmental cold and emotional austerity. Descriptions of her palace, the environment she has shaped out of snow and wind, inform our understanding of the Queen herself:

> The biggest room stretched for miles, and all the rooms were so large, so empty, so shiny, and so icily cold. No one ever had any fun there—there wasn't even a dance for the polar bears. […] There were no card games with paw-smacking and backslapping or even a ladies' coffee klatch for the white foxes.
>
> *(Andersen 2009, 200)*

The Snow Queen is a feminine figure defined by lack. As Andersen's whimsical multi-species counterexamples indicate, this includes the lack of interpersonal and domestic warmth, qualities essential to the performance of the archetypal 'good' wife, mother or hostess. The emptiness of the Snow Queen's domain is also indicative of her physical barrenness, her failure to reproduce biologically. Though attractive to Kai ('delicate and beautiful'), she is at times described as being made of 'blinding, glimmering ice' and is by extension repellent to touch or even sight (Andersen 2009, 172). She is seductive but ultimately withholds romantic and emotional satisfaction from Kai, offering neither sex nor maternal care, a dereliction of heteronormative gender roles that is at times entangled with lethal physical danger: 'I shall kiss you to death' she teases as she takes him to the ice palace.

The Snow Queen is placed in contrast to Gerda, whose journey exhibits the values of girlish innocence and sometimes maternal affection. Gerda's goodness is characterised by protectiveness, tenderness and generosity towards Kai, which manifests throughout much of the story as feminine sacrifice of her own possessions, happiness and even safety. The Snow Queen's arctic domain of intellectualism and non-productive companionship appears opposite the fecund southern climes and roses to which Kai and Gerda emerge for their happy ending in the emotional warmth of domesticity. 'And it was summer' the story concludes, 'warm, wonderful summer' (Andersen 2009, 204).

Like other fairytales, 'The Snow Queen' gestures to didacticism from the start: 'When we're at the end of the story, we'll know more than we do now' the narrator proclaims. Yet, its narrative complexity and tonal playfulness make morals difficult to pin down, and the story resists heroism and villainy. Andersen's Snow Queen is not the devil (who creates the offending mirror), nor even his accomplice; she is an independent entity of great power who finds some affinity with Kai in his altered state. She has a seemingly amoral constitution, driven by aesthetics and whims that seem strange, at a distance—both in terms of affective valance and spatial scale—from motives that animate the other characters. For example, in the final act, she leaves Kai for a trip to see some volcanoes, explaining: 'I want to take a look at the big black pots [...] I want to give them a coat of whitewash. They need it, and it will look good after the lemons and grapes have been harvested' (Andersen 2009, 201). For all she lacks, the Snow Queen does possess intellect and engages Kai in puzzles, but this companionship is implicitly monstrous, attracting the boy only because his sensibilities have been inverted by the mirror. Their perverse attraction is one instance of the queer subtext often detected in readings of Andersen's work and autobiography. Anderson never married or had children and authorial choices such as the child-like resolution of Gerda and Kai's love story have been interpreted as suggestions of sexual repression (Zipes 2014, 10).

Conversely, we might choose to read Andersen as representing a range of relationships and gender possibilities in which the central plot is only an artefact of narrative cohesion and not an endorsement of a particular value hierarchy. To my mind, Andersen's story reads as a tale in which power and agency are dispersed across a remarkably broad array of charismatic characters. For all her adherence to normative feminine qualities, Gerda is the hero who sets out on a daring journey to free Kai, and whose agency propels most of the plot. This disrupts a pattern that anthropologist Sherry Ortner identified as core to the construction and maintenance of gender across Grimms' fairytales, namely that feminine characters are consistently 'made to be passive' and renounce agency; women characters who exhibit agency (generally witches or step relatives) are punished for their wilfulness (1997, 9). Andersen's story teems with exceptions to this characteristic of earlier European folklore.

As many post-humanists have observed, the issue of agency (or 'animacy', a more general term Mel Y. Chen develops to describe the capacity of other-than-humans to affect the world with or without intentionality [2012]) is central to the construction and maintenance of not only gender but also the categories of the human and non-human. In parallel to a traditional femininity that is shaped by the repression of women characters, the construction and enactment of the 'natural' generally depends upon a denial and repression of agency among entities deemed to be non-human (e.g., animals, plants, or ice). This is a connection that Andersen's text seems to intuit in that it playfully subverts both norms, often at the same time. The agential beings Gerda encounters throughout the journey, both helpers and adversaries, are almost entirely feminine and/or non-human: she is given shelter by two old women who each live alone, is kidnapped and then set free by a brash, devil-may-care bandit girl, is guided by a pair of ravens, speaks to a garden of flowers, bargains with a river and, ultimately, depends on the expert knowledge and goodness of a talking reindeer to reach Kai. The Snow Queen is described as the head of a bee colony, and snowflakes as white bees who serve her. Gerda encounters the Queen's guard, 'a whole regiment of snowflakes' that 'were alive' and animalic as well, shaped 'like big ugly hedgehogs' and 'small chubby bears' (Andersen 2009, 198). Most of these feminine and non-human entities leave Gerda's narrative for new exploits, their wilfulness unpunished and intact.

There are climate implications in the complex, diverse animism of Andersen's storyworld, which defies normative hierarchies of gender, species and liveness. Climate change can be understood as a symptom of the global systematic denigration and long-term exploitation of people and beings designated as 'non-humans', collectively as 'nature'. In contrast to the hegemonic Western rendering of non-humans as inert and fungible material resources, the Snow Queen is both the ruler of snow and ice and the personification of these elements. In other words, Andersen's story positions an entity normatively designated as non-human and non-agential as an entity with intentionality and extraordinary power.

Close reading of Andersen's ending even suggests that it is the ice itself that benevolently frees Kai from an excess of cold and concomitant logic. In the palace, Kai became obsessed by 'an ice puzzle of the mind', and is in effect a prisoner of his own fixation. Kai 'had been struggling to shape the pieces of ice into a word—*eternity*' and the Snow Queen offered: '"If you can make that design for me, you'll be your own master, and I'll give you the entire world and a pair of new skates". But he could not do it' (Andersen 2009, 201). When Gerda and Kai reunite, their happy tears melt the mirror splinters away, an event 'so wonderful that even the pieces of ice danced happily around. When they got tired, they lay down, forming just the word that the Snow Queen had asked' (Andersen 2009, 201). The story's resolution celebrates self-mastery as the key to freedom, including the ability to navigate the environment (ice skates) and to escape an inhospitable climate (a return journey to their village, marked by birdsongs and green sprouts). However, this mastery and freedom is a gift of the ice, and by extension, of the Snow Queen herself: 'his freedom was written on the floor in shining pieces of ice' (Andersen 2009, 201). Climatic disequilibrium, the story's central conflict, is resolved through the ice's agency which transforms Kai (rather than Kai's manipulation of the ice.) Andersen's original story suggests that humans should not regard the non-human world as an antagonist or a problem to be solved and must instead give way to the agencies of the non-human world to attain their own freedom.

Yet, despite the nuances and ambiguities of Andersen's text, the Snow Queen circulates as a widely recognised and basically derogatory gender trope. Andersen's character did become villainous in retellings. A popular 1956 Russian cartoon adaptation of Andersen's story featured only one major change, which was to make the Snow Queen vengeful and cruel. In this telling, she was personally motivated by anger with Kai (because he taunted her power) and therefore she intentionally pierced Kai with the mirror. Disney also imagined the Snow Queen as the story's villain across repeated stymied attempts to adapt the story, beginning in the 1930s and continuing well into *Frozen*'s development (Shaw-Williams 2020). These adaptations acquiesce to deeply engrained assumptions of gendered value. As a symbol of women who neglect to fulfil the most culturally naturalised (as well as economically prescribed and socially prized) feminine roles within heteronormative patriarchal societies, namely emotional caretaking and biological motherhood, the Snow Queen is anathema.

The devaluation of the feminine in association with the snowy Arctic is also a devaluation of the snowy Arctic in association with the feminine. The personification of the Far North as a villainous Snow Queen nurtures a widespread presumption that the Arctic, likewise, is under-productive and unnaturally resistant to domestication. Like the Snow Queen, the land of the Far North becomes an aberration defined by lack, the failure to (re)produce and support civilisation in the heteropatriarchal, sedentary, agricultural and capitalist mode that came to dominance in Western Europe and through European colonization.[3]

This misperception overlooks the human societies and ecological communities that thrive in the Far North, not despite snow and ice but in ways that rely upon their unique capacities. Across national contexts, reductive and entangled interpretations of the Snow Queen and the Arctic as entities that fail to meet the 'natural' (read 'naturalised') teleology of feminine bodies also serve to naturalise the widespread devaluation of polar places. Demonisation of the Snow Queen both mirrors and supports the ongoing mass violence of industry-driven climate change and the rapid, exogenous transformations of the Far North as a place supposedly lacking inherent value.

Of course, the movie *Frozen* famously disrupts several of these dynamics: it challenges the demonisation of the Snow Queen and centres a story of sisterly (rather than romantic) love. *Frozen* essentially recasts the fairytale's best friends as two siblings, the Princesses of Arendelle: Anna is a close equivalent to Gerda, while Elsa is an amalgam of the Snow Queen and Kai. Elsa is both a force of nature who can create ice and snow with her hands and a person alienated from her own emotions (with which her powers are intertwined). As a child, Elsa accidentally hurts Anna in an outburst of fear and cold and her father trains her in obsessive self-control with the mantra 'conceal, don't feel'. The sisters are emotionally estranged until their parents die, and Elsa is crowned Queen. At the coronation celebration, Elsa becomes overwhelmed by Anna's impulsive engagement with a neighbouring prince and causes the ballroom to freeze, revealing her long-concealed powers. She then retreats to the North Mountain and, in show-stopping fashion, resolves to 'let it go', expressing her powers and her feelings with unbridled self-confidence, lyrically embracing freedom and faith in progress. These overflowing emotions create an ice palace and trigger rapid climate change that threatens to devastate Arendelle with endless freezing temperatures. Anna (like Gerda) then undertakes a brave journey to save her beloved (Elsa) from the Snow Queen (also Elsa).

Andersen's story finds a happy resolution through Gerda's self-sacrificing love, Kai's emotive freedom and the animacy of the ice itself. *Frozen's* resolution similarly pivots on Anna's self-sacrificing love but also, crucially, on Elsa's emotive moderation and self-mastery, modelling a kind of techno-futurist climate control. Elsa learns to regulate her emotions and meticulously shape her inner and outer worlds; ice, resultantly, comes to resemble plastic in its frivolous, consumer-oriented functions. By *Frozen II*, for example, Elsa's superpowers have been used to manufacture ice toys, decorate tablescapes at parties with ice and shape mobile micro-climates, accommodating the personalised climate needs of her snowman friend, Olaf, throughout the fall.

Other scholars have read *Frozen* as a pedagogically dissatisfactory climate narrative, arguing that the dramatic denouement of Elsa's self-discovery valorises neoliberal individualism 'and "Self-love" as the solution to the very problems and conditions it engenders' through socioeconomic inequities (Blouin and Shipley 2002, 190). The film's celebration of individual success in capitalist terrains is clear: Arendelle's climate crisis is rendered in economic terms, including an oversupply of ice as a commodity, and the film's lyricist, Kristen Anderson-Lopez, cites American self-help authors (including Brené Brown) as key influences on her songwriting (Buck and Lee 2013; ABC News 2019). The collapse of Kai and the Snow Queen into one character largely transforms a story of discord and harmony across difference into a journey of self-optimization. Blouin and Shipley conclude that while the film 'positions human innovation as the means to curb human-caused destruction, the real solution,' that is, what is required actually to combat anthropogenic climate change, 'manifests in moments of mutuality and self-sacrifice' (2002, 200). It seems easy to sympathise with these scholars' guiding reverence for nature and their indictment of *Frozen*'s (clearly evident) embrace of neoliberal values. Yet, this discussion does not specify why recognition of mutuality or self-sacrifice are the interventions that contemporary climate change demands, materially or politically; these are essentially ethical presuppositions (ones that are common to many eco-critics) against which *Frozen* is analysed.

Additionally, no discussion I've encountered acknowledges the actual innovations implicated in Elsa's heroic attainment of climate control: technologies of geoengineering. Geoengineering is a term that names any effort to deliberately alter the climate, and while it is mostly a topic of theoretical discussion, geoengineering methods are being increasingly tested by academic, industrial and government institutions (NPR 2024). Solar geoengineering proposes methods to increase the reflectivity of the stratosphere so that less sunlight reaches earth, for example, by injecting aerosols into the atmosphere and thereby changing the composition of clouds (Buck 2019). Physicist John Latham proposed such an approach to 'control global warming' back in 1990. Yet many in the climate science sector are extremely reticent to discuss such possibilities while attempting the daunting work of mitigating global carbon emissions; some even suggest a moratorium on solar geoengineering research. Holly Jean Buck demonstrates the tone with which solar geoengineering is often discussed:

> It's been skulking in the shadows of climate policy for a decade and haunting science for longer than that [...] It is unlikely you imagine solar geoengineering would be a serious topic of discussion, because it sounds too crazy [...] Indeed, it *is* a drastic idea.
>
> *(2019, 6, emphasis in original)*

I would offer that generations of children who grew up with climate crisis fairytales that resolve in fine-tuned, indefinite climate control may not find the notion so monstrous nor drastic. Solar geoengineering is rarely mentioned without the asterisk of inherited wisdoms: about hubris, about the ethical good of sacrifice and about the limitations of technological fixes (of which there are many in this instance, including unpredictable global ramifications to agriculture and the need for continuous upkeep of geoengineering to prevent 'termination shock') (Harvey 2023). I repeat these cautions and simultaneously, am compelled to observe that in the decades since Latham's proposal, catastrophically little progress has been made to keep projected warming at manageable levels through the reduction of carbon emissions. Recent estimates suggest that the first day when the Arctic will be effectively 'ice-free' will occur within the next decade, even in the (currently politically foreclosed) scenario of extreme emissions reductions (Jahn et al. 2024) and the complex effects of this will cascade globally, over generations.

My own worldviews incline me to wish our systems and laws might avert further climate violence through the logics of Andersen's *Snow Queen*. I want to affirm that respect for the agency of non-human beings (still) produces reciprocities and sustains relational and ontological diversity; or, that emotional and spiritual experience have value above and beyond capitalist accumulation and growth. However, such suggestions seem increasingly remote from our lived sociopolitical and geophysical climate—fantastical, even, given the abandonment of global climate accords by key nations. *Frozen*'s vision of unilateral, geographically indifferent, human-centric climate control in the form of geoengineering is by no means a sustainable panacea or even (to our knowledge) a feasible short-term solution for rapidly accruing climate challenges. This is particularly true because the singular hero model (echoed by the self-authorised geoengineering startups experimenting with aerosols as I write) does nothing to address climate's fundamental demand for global coordination and cooperation (NPR 2024). The addition of intentional geoengineering (to the *de facto* experiment in anthropogenic planetary warming) might only expand the as-yet endless appetite for energy usage, which has grown in parallel with renewable power infrastructures (Perkins 2023). This is not an endorsement, but an acknowledgement that in our present conditions, Elsa's brand of geophysical magic appears to be swinging from the realm of the 'drastic' towards the increasingly imaginable among measures to sustain a survivable climate. Absolute climate control may be a fairytale, but it may also be an inevitable ambition of a voracious neoliberal system that conditions and contorts feminized, racialized, and other-than-human bodies to accommodate limitless desires for economic productivity.

Notes

1 Polar bears exclusively inhabit the greater arctic region, and their images are routinely enlisted as evidence of a warming climate, and as visual symbols for the potential environmental costs of unmitigated climate change. In 2006, the landmark climate awareness film *An Inconvenient Truth* featured an animation of a polar bear struggling to find solid sea ice on which to rest or hunt while Al Gore described how polar bears had been observed swimming extraordinarily long distances and sometimes drowning in the warming Arctic. Dominate climate communication's metonymic

reliance on polar bears and changes to arctic ecosystems has persisted in the decades since that film's release. As of 5 March 2025, the first 100 results of a Google Image Search of the term 'climate change' featured landscapes and just two animal species, humans and polar bears, represented in roughly equal measure.

2 The location of the Snow Queen's palace is sometimes referred to as 'fearsome, ice-cold Finnmark' (Andersen 198) and sometimes as Spitsbergen (193), an Arctic Norwegian archipelago also known as Svalbard.

3 The gendered logics I describe here resonate with dominant colonial sensibilities that Jen Rose Smith terms 'temperate-normativity' (2021, 1).

References

ABC News. 2019. '"Ep. 2: Return to Arendelle" and "Ep. 7: Legacy, Influence and Impact."' *Inside Frozen 2*. 25 December. [Podcast]. https://abcaudio.com/podcasts/inside-frozen-2/, accessed 4 March 2025.

Andersen, Hans Christian. 2009. *The Stories of Hans Christian Andersen: A New Translation from the Danish*. Translated by D.C. Frank and J.C. Frank. Durham, NC: Duke University Press.

Blouin, Michael, and Morgan Shipley. 2002. '"How Does Nature Deify Us": Disney, Climate Change, and the Worship of the Self.' *CR: The New Centennial Review* 22 (2), 161–203.

Buck, Holly Jean. 2019. *After Geoengineering: Climate Tragedy, Repair, and Restoration*. London: Verso.

Buck, Chris and Jennifer Lee, dir. 2013. *Frozen*. Walt Disney Studios Motion Pictures.

Cama, Timothy. 2017. 'Emails: Disney Annoyed by Obama Push to Use "Frozen" Brand.' *The Hill*, 12 December. https://thehill.com/policy/energy-environment/364370-emails-disney-annoyed-by-obama-push-to-use-frozen-brand/, accessed 27 February 2025.

Chen, Mel Y. 2012. *Animacies: Biopolitics, Racial Mattering, and Queer Affect*. Durham, NC: Duke UP.

Harvey, Fiona. 2023. 'Experts Call for Global Moratorium on Efforts to Geoengineer Climate.' *The Guardian*, 14 September. https://www.theguardian.com/environment/2023/sep/14/experts-call-for-global-moratorium-on-efforts-to-geoengineer-climate, accessed 27 February 2025.

Hewes, Jane, P. Whitty, Brittany Aamot, Erica Schaly, Jennifer Sibbald, and Kayla Ursuliak. 2016. 'Unfreezing Disney's Frozen through Playful and Intentional Co-Authoring/Co-Playing.' *Canadian Journal of Education* 39 (3), 1–25.

Holbek, Bengt. 2002. 'Hans Christian Andersen's Use of Folktales.' In Anna Chaudhri and Hilda Davidson, eds, *A Companion to the Fairy Tale*. Woodbridge: Boydell & Brewer, 149–58.

Jahn, Alexandra, Marika Holland, and Jennifer Kay. 2024. 'Projections of an Ice-Free Arctic Ocean.' *Nature Reviews Earth & Environment* 5 (3), 164–76. https://doi.org/10.1038/s43017-023-00515-9.

Konnikova, Maria. 2024. 'How 'Frozen' Took Over the World.' *The New Yorker*, 25 June. https://www.newyorker.com/science/maria-konnikova/how-frozen-took-over-the-world, accessed 27 February 2025.

Latham, John. 1990. 'Control of Global Warming?' *Nature* 347 (6291, September), 339–40. https://doi.org/10.1038/347339b0.

NPR. 2024. 'The Sunday Story: Startups Want to Cool Earth by Reflecting Sunlight.' *Up First*, 28 April. https://www.npr.org/2024/04/28/1198912483/solar-geoengineering-few-rules-big-risks, accessed 27 February 2025.

Ortner, Sherry. 1997. *Making Gender: The Politics and Erotics of Culture*. Boston, MA: Beacon Press.

Perkins, Robert and Henry Edwardes-Evans. 2023. 'Fossil Fuels "Stubbornly" Dominating Global Energy despite Surge in Renewables,' *S&P Global*, 26 June. https://www.spglobal.com/commodityinsights/en/market-insights/latest-news/oil/062623-fossil-fuels-stubbornly-dominating-global-energy-despite-surge-in-renewables-energy-institute.

Rantanen, M., A.Y. Karpechko, A. Lipponen, K. Nordling, O. Hyvärinen, K. Ruosteenoja, T. Vihma, and A. Laaksonen. 2022. "The Arctic Has Warmed Nearly Four Times Faster Than the Globe since 1979." *Communications Earth & Environment* 3 (1), 1–10. https://doi.org/10.1038/s43247-022-00498-3.

Shaw-Williams, Hannah. 2020. 'Frozen: Every Failed Disney Attempt to Adapt the Snow Queen.' *ScreenRant*, 27 May. https://screenrant.com/frozen-movie-disney-snow-queen-adaptation-plans-fail/, accessed 27 February 2025.

Smith, Jen Rose. 2021. "Exceeding Beringia': Upending Universal Human Events and Wayward Transits in Arctic Spaces.' *Environment and Planning D: Society and Space* 39 (1), 158–75. https://doi.org/10.1177/0263775820950745.

Zipes, Jack. 2014. *Hans Christian Andersen: The Misunderstood Storyteller*. New York: Routledge.

22

'IT DOESN'T WORK. OF COURSE IT DOESN'T WORK. BUT IT DOESN'T MATTER THAT IT DOESN'T WORK'

David Finnigan's Dramaturgy of Errors

Milo Harries

David Finnigan's collaborators often approach his practice expecting it to work in a certain way. Finnigan is a writer and theatremaker of white settler heritage from Ngunnawal country, Australia, who situates his climate-engaged practice 'at the intersection of science and art' (Finnigan 2025). His work, which ranges from plays to interactive performances and games, is often produced in collaboration with climate scientists or based on scientific research. This positioning means that Finnigan's practice is at times pre-emptively understood within a 'problem-solution' frame (Woynarski 2020, 2), supposing a direct, communicative theory of change:

> talking with scientists, a lot of the time there's this moment where they'll be like—so you're interested in my science—maybe you could use your art, and your art could get through to people because art connects with people in a way that science doesn't, and that will […] turn the dial.
>
> *(Finnigan 2021)*

Finnigan's comment here helps to clarify the model to which his work is often expected to conform. Using art to 'turn the dial' recalls what Mike Hulme refers to as the 'deficit model' of climate communications: a relationship of 'linear flow', predicated on the sentiment that 'we are the ones with the truth, the prophets who can see the future, and it's these people that we have to convince' (Harrabin et al. 2011, 85). This model is legible in the logic of the early work of British director Katie Mitchell, whose ongoing engagement with the climate crisis began in response to a request from the data scientist Stephen Emmott, who asked 'whether theatre could help communicate what was happening in

DOI: 10.4324/9781003584445-29

any way' (Lease 2020, 257). Subsequently collaborating with Emmott on the lecture-performance *Ten Billion*, Mitchell attempted to use the theatre as a vehicle for communication, '[placing] the scientist onstage, to explain the data with clarity and objectivity from a position of authority' (Macmillan and Rapley 2016, 231). In this mode of performance, then, success is identified with persuasion—convincing audiences of the reality of the crisis by deploying the lucid authority of the stage.

This model, however, has come under increasing pressure within Anglophone 'climate theatre' in the last 10–15 years. As Catherine Love observes, for example, Katie Mitchell's more recent practice has pivoted 'from facts to feelings' (Love 2020), at times overtly challenging her earlier approach: most obviously, a shift in 2023 saw Mitchell direct a Barbican (UK) production of Miranda Rose Hall's *A Play for the Living in a Time of Extinction*, in which a dramaturg character named Naomi recants her erstwhile desire to use 'KNOWLEDGE AND INFORMATION' to '[BEAT] AN AUDIENCE INTO SUBMISSION, / into ACCEPTING the ERROR OF THEIR WAYS' (Hall 2023 [2021], 16, original emphasis). This evolution in Mitchell's practice is indicative of broader trends: *A Play for the Living* is only one of a number of recent pieces (including Alannah Mitchell's *Sea Sick* and Fehinti Balogun's *Can I Live?*) that explicitly disavow the authority of an onstage scientist, instead centring artists who are ordinary, entangled and even overwhelmed.

In keeping with this pattern, and in contrast to the expectations often brought to his work, Finnigan's practice begins from a far less authoritative approach to his role as an artist, and a very different understanding of what it might mean for a performance to succeed. Finnigan's work, in fact, often works by not working, taking failure as an opportunity to re-imagine existing narratives and frames. Indeed, 'failure', as it is discussed in this chapter, is understood more as a broadening of vision, revealing the contexts and contingencies that surround specific understandings of success (Bailes 2011, 34). In this regard—although their thinking derives from very different contexts—Finnigan's work echoes Jack Halberstam's argument that failure can be a way to '[exceed] the frameworks with which one enters', creating novel paradigms and relationships by reaching beyond conventional understandings of success (2011, 12). In this chapter, I explore the potential of this capability to demonstrate some of the stakes of the broader cultural shift outlined above—concluding that fallible artists can create opportunities that more authoritative approaches might foreclose.

Pilot Episode

Finnigan engages directly with persuasive models of science-based 'climate theatre' in *Pilot Episode*, the first instalment in his ongoing series *You're Safe*. *Pilot Episode* is performed by Finnigan as a performance lecture, with support from musician Reuben Ingall, and begins by promising the kind of 'sugarcoated pill' that audiences for 'climate theatre' have come to expect (Chaudhuri 2017, para. 6). After a brief prologue, Finnigan tells the audience about his experience at a recent networking event, during which he failed to convince a financial advisor named Vishnu to take action on climate change. This incident, Finnigan explains, has motivated him to learn how to communicate more effectively around climate action. 'The next time I meet a financial advisor at a networking event', Finnigan announces, 'I'm going to be ready. I will be persuasive, engaging, light, funny, informative. I'm going to have all the answers ready to go' (Finnigan 2019b, 4). This intention supplies the premise of the performance: Finnigan informs the audience that he will take the audience with him as he recounts a series of conversations with climate scientists through which he hopes to return triumphantly to his earlier conversation. At the end of the piece, he will repeat this encounter, addressing the audience as if they were Vishnu: 'only this time', as he later insists, he will 'win [Vishnu] over with facts and charisma' (Finnigan 2019b, 19).

Finnigan thus puts *Pilot Episode* on a redemptive arc, offering a test case for communicative climate theatre's ability to persuade and inspire. He in fact overtly identifies the performance in these terms, echoing Duncan Macmillan's sense that a key part of an artist's role in this kind of work is to 'make the data dramatically compelling to listen to' by telling the audience that he wants to make 'really abstract' material 'flare in [their] chest' (Merritt 2014, para. 5; Finnigan 2019b, 5). Most of the piece is correspondingly structured as a series of educational episodes, with Finnigan drawing on conventional science-writing techniques to make his material 'exciting and explosive' by translating abstract concepts into accessible human scale (Finnigan 2019b, 5). Narrating a conversation with Earth system scientist Will Steffen, for example, Finnigan explains the Great Acceleration in terms of the average farmed chicken's increase in weight, visualising this change by filling a plastic bag with water (Finnigan 2019b, 8). He even comments on his tactics as he does so: 'if you want a tangible example, Vishnu', he exclaims, 'here's a goddamn giant chicken' (Finnigan 2019b, 9). By not only deploying but also foregrounding these tactics, Finnigan emphasises the show's identity as a 'sugarcoated pill' (Chaudhuri 2017, para. 6), anticipating a successful culminating act of persuasive communication.

As *Pilot Episode* arrives at its conclusion, however, it stages its own failure to 'turn the dial' (Finnigan 2021), with Finnigan's carefully gathered vignettes refusing to cohere into an overarching narrative. Rather than having 'all the answers ready to go' (Finnigan 2019b, 4), Finnigan can only deliver his illuminating examples in a confusion of fragments, his thinking jumping from place to place in a frustrated attempt to make sense of the crisis as a whole:

> trillions upon trillions of fossilised chicken bones buried in
> /

> a cheetah is designed to operate
> /
> more full of plastic than any human in any generation before.
>
> *(Finnigan 2019b, 23)*

By one standard, then, *Pilot Episode* ends in failure, with Finnigan finding himself incapable of performing the task that he set out to accomplish. Crucially, however, *Pilot Episode* treats this failure as a form of escape, challenging the framework with which its story began. Finnigan's failure, that is, turns out to be stakeless: 'And it doesn't work', Finnigan reflects, 'Of course it doesn't work. But it doesn't matter that it doesn't work' (Finnigan 2019b, 24). Rather than lamenting his failure to communicate as a personal shortcoming, as he did after his first conversation with Vishnu (Finnigan 2019b, 3), Finnigan rejects his task as an impossible challenge, based on an unreasonable expectation:

> I've spent 18 months trying to capture this whole global crisis in a few snappy sentences and images. But it keeps sliding out of my grip.
>
> [...]
>
> You can't turn this crisis into a story with a simple moral and a call to action. Not in a conversation in a bar with someone you've just met, not in a theatre with a group of strangers.
>
> *(Finnigan 2019b, 25)*

Vitally, moreover, Finnigan treats this realisation as an opportunity to re-negotiate the relationships from which that expectation emerged. Countering Hulme's division between 'the ones with the truth' and the people they have to convince (Harrabin et al. 2011, 85), Finnigan insists instead on the audience's own responsibility for how they feel, and what they do: 'I'm not responsible for how you feel about this stuff', Finnigan insists; '[i]t's not up to me to make anyone hopeful or convince anyone to take action' (Finnigan 2019b, 25). This assertion allows *Pilot Episode* to escape from the 'linear flow' of its opening frame (Harrabin et al. 2011, 85) and propose instead a collaborative relationship predicated on individual responsibility: 'You give yourself hope', Finnigan suggests, 'You make yourself useful' (Finnigan 2019b, 25). Finnigan's failure thus allows *Pilot Episode* to act as a starting point: a chance, in Alan Read's terms, 'to reassemble the social on renewed terms, to begin again' (2008, 6). These are the terms, in fact, in which the piece frames its ending: 'So I'll ask you', Finnigan concludes, 'what's useful? Where do we begin?' (Finnigan 2019b, 25).

Pilot Episode thus helps to highlight one potential in a more ordinary, fallible understanding of an artist's role: rather than authoritatively delivering information, an openness to error opens space for the unexpected, allowing novel possibilities to emerge. This is a more fluid, more situated and less certain position, rejecting pre-determined answers whilst insisting on responsiveness: it is, in Emily Ogden's terms, 'not the defensiveness of willful ignorance but the defenselessness of not knowing yet' (2022, 6). In its vulnerability, however, what this position allows is the chance for something new. I will therefore now turn to Finnigan's earlier play *Kill Climate Deniers* to demonstrate what that novelty can enable and illustrate the forms of success that failure can achieve—showing how Finnigan's rethinking of his mistakes can re-structure destructive social forms.

Kill Climate Deniers

Whilst *Pilot Episode* ends by linking error and emergence, this connection is where Finnigan's earlier play *Kill Climate Deniers* begins. As *Kill Climate Deniers*' title suggests, its central subject is the cultural politics of climate denial, and the vicious antagonism of the climate 'debate'. Belying the title's tone, however, the play does not set out to fight in this ongoing culture war, or to correct its antagonists' conceptual frames. Instead, *Kill Climate Deniers* opens with an author-character named finig addressing the audience directly, saying: 'First thing I got wrong was the title' (Finnigan 2018, 11). This self-critical beginning establishes an authorial identity that is not authoritative or antagonistic but error-prone and calm. It frames the play, in part, as a reflection on its own production: 'You want to call your play something fun, something playful, something catchy', finig muses, 'but sometimes you get it wrong, you get it wrong, you get it wrong, you get it wrong' (Finnigan 2018, 11). Throughout *Kill Climate Deniers*, finig correspondingly appears at intervals to recount and re-consider the events leading to the play's production, exploring what can be learned from Finnigan's mistakes.

This process of emergence seeks to escape from the static antagonisms that had shaped *Kill Climate Deniers*' journey to the stage. When the play was awarded public funding in 2014, its title caused an outcry from the political right, both in Australia and online. Claiming that *Kill Climate Deniers* represented 'the hatred of the Left', for example, an InfoWars presenter portrayed it as a move in a violent culture war: '[t]he Left is honestly a death cult [...]. There's a reason why they have a play called *Kill Climate Deniers*. That's what they want to do. They want to kill you' (Finnigan 2021). Threats were made to the playwright, company and cast, forcing a postponement that meant *Kill Climate Deniers* was only first performed as a conventional play in 2018, after Finnigan won the 2017 Griffin Award (although in the interim, Finnigan had released versions of the play as an ebook, dance party, music video, concept album and walking tour [Finnigan 2021]). The story of the play's production, then, speaks to a landscape of aggressive polarisation between entrenched, opposed, uncompromising groups.

Part of *Kill Climate Deniers*' work is to illustrate the destructiveness of these frozen social relationships. Alongside the scenes involving the author-character finig, *Kill Climate Deniers* offers its audience another set of scenes, translating the film *Die Hard* into an eco-terrorist siege of Australia's Parliament House, with Environment Minister Gwen Malkin in the role of John McClane. This story is driven by groups whose identities cannot be reconciled, driving an increasingly violent antagonism: Malkin's '[y]ou fuck with Australia, I will fuck you up' (Finnigan 2018, 120), for example, collides head-on with lead terrorist Catch's '[w]ho's got bad values and wants to die?' (Finnigan 2018, 130). Although these action-movie scenes are light-hearted, they offer an analysis of the denial 'debate' by staging 'two powerful and opposing political forces, pushed, by the extremity of the situation and the immovability of their positions, into a lethal death spiral' (Hobba 2018, 7). The action-movie arc, in other words, offers a critique of the denial debate's frozen relationships through its form—with the requirements of the genre implicitly arguing that this struggle can only end, as *Die Hard* does, with the total elimination of one side.

Kill Climate Deniers thus centres what Sarah Dillon and Claire Craig describe as '[t]he increasing narrative lock-in and associated polarisation of collective identities in relation to climate change' (2021, 74). Collective identities, Dillon and Craig argue, are created and consolidated by stories, which produce and reproduce prevalent social groups and individuals' senses of their positions within or outside them (2021, 57–88). The term 'narrative lock-in' thus points to the limited social identities that are available within an unchanging narrative frame (74)—including those adopted and embodied by action-movie hero Gwen Malkin, and the terrorists that she picks off one by one. When these stories are rigidly held ('locked in'), so are the collective identities they imply—leading to fixed, antagonistic and even potentially violent understandings of 'us' and 'them'. This sense of inimical, rigidly held and reciprocally defined identities is evident in the online comments Finnigan gathers in the text of the play:

> dirt worshipers [*sic*] are pu$$ies. bring it on.
>
> Are you prepared though, to kill them before they kill us? I certainly am, and I'm devoid of remorse for doing so.
>
> […]
>
> I'm a denier. Now come and kill me. I'll fertilize the earth with your rotting corpses. Just giving back to mother earth.
>
> […]
>
> Come try and kill us, Liberals.
>
> *(Finnigan 2018, 77–80, original formatting removed)*

Alongside its alertness to this polarisation, however, *Kill Climate Deniers* also offers its audience a means of escape. By focussing on Finnigan's mistakes, finig establishes what Theresa May elsewhere identifies as one of theatre's core potentials: 'a kind of *bracketed time* in which we set aside both the demands and the differences that tear at democratic values' (2021, 280, original emphasis). Reflecting on a successful ambush that Finnigan laid for local conservative politician Brendan Smyth, for example, finig urges the audience to 'listen again' to Smyth's response and his reasons for objecting to the play's title: 'Was he right?', finig asks: 'Is it okay to upset and offend people with a hyperbolic title like this? Does the end justify the means?' (Finnigan 2018, 26–27). finig's openness to error thus opens space to reflect, re-think and renew—re-telling the story of the play's production to escape the denial debate's locked-in narrative frames.

Rather than recommencing *Kill Climate Deniers*' hostilities, then, the finig scenes identify Finnigan's mistakes as pathways to reconciliation—using an openness to error as an opportunity for this story to be re-told. This re-narration of *Kill Climate Deniers*' past, that is, allows finig to re-formulate the identities that its antagonisms implied. As the play nears its ending, deniers are no longer understood as being blind to the science, but instead clear-sighted about its implications: '[y]ou know what it means if it's true', finig reflects, '[s]o you can't allow it to be true' (Finnigan 2018, 129). Those who supposedly accept the science, meanwhile, are suspected of their own forms of denial, with both groups '[acting] as if the world's not turning upside down around us' (Finnigan 2018, 129): 'everything you care about is also everything I care about', finig reflects, 'but I haven't come to terms with the fact that it's gonna go' (Finnigan 2018, 139). Admitting to having once 'hoped [climate deniers] would die', meanwhile, finig now accepts the idea that deniers are 'always going to be here', defying the action-movie's logic of elimination to argue that the 'real fight' is a shared struggle to negotiate loss and change (Finnigan 2018, 139).

Through this reworking of *Kill Climate Deniers*' initial antagonisms, finig thus opens a path to identities and futures that transcend the play's original frames. This iterative reflection allows the play, in closing, to reject the blunt and binary opposition out of which it emerged, affirming instead an inclusive principle from which to 'begin again' (Read 2008, 6). The play ends by returning to an ethic of pluralism, a quotation from the Bible not assigned to a character but simply printed twice, underlined, as the play's epigraph and as its final words:

> whoever is not against us is for us
> whoever is not against us is for us
>
> *(Finnigan 2018, 12, 141)*[1]

By opening itself to Finnigan's errors, then, *Kill Climate Deniers* re-tells its own story, creating new identities that are not predicated on the elimination of the other side. And in doing so, *Kill Climate Deniers* precedes *Pilot Episode* in rejecting an idea of the artist as someone with 'all the answers ready to go' (Finnigan 2019b, 4)—proposing instead an artist who is always ready to go in search of the answers, regardless of the enmities in which they are embroiled. Like *Pilot Episode*, therefore, *Kill Climate Deniers* argues for the value of an artist imagined not as a prophet but a project: no longer authoritative, or aggressive, but receptive, reflexive and unsure. This kind of artist, the play suggests, can find ways forward through failure and allow more generative forms of relationship to emerge.

Note

1 In the printed edition of the text (Finnigan 2019a), this phrase appears as the epigraph (12) but does not return at the end of the play.

References

Bailes, Sara Jane. 2011. *Performance Theatre and the Poetics of Failure*. Abingdon: Routledge.

Chaudhuri, Una. 2017. 'Climate Lens: Birth of a Post-Nation!' *HowlRound*, 21 April. https://howlround.com/climate-lens, accessed 21 March 2025.

Dillon, Sarah and Claire Craig. 2021. *Storylistening: Narrative Evidence and Public Reasoning*. Abingdon: Routledge.

Finnigan, David. 2018. *Kill Climate Deniers*. Digital ed. https://davidfinig.com/shop.

Finnigan, David. 2019a. *Kill Climate Deniers*. Print ed. London: Oberon.

Finnigan, David. 2019b. *You're Safe Til 2024: Pilot Episode*. Unpublished manuscript in possession of the author.

Finnigan, David. 2021. 'Does Someone Deserve to Die for This? Theatre, Climate, and Global Change.' Public conversation with the author conducted on Zoom for a University of Cambridge Drama and Performance Graduate Seminar, 4 November 2021.

Finnigan, David. 2025. 'About.' *David Finnigan*. https://davidfinig.com/about/, accessed 1 May 2025.

Halberstam, Judith [Jack]. 2011. *The Queer Art of Failure*. Durham, NC: Duke University Press.

Hall, Miranda Rose. 2023 [2021]. *A Play for the Living in a Time of Extinction*. London: Nick Hern Books.

Harrabin, Roger, Mike Hulme, Ruth Little, Oliver Morton, and Carolyn Steel. 2011. 'Futures.' In Robert Butler, Eleanor Margolies, Joe Smith and Renata Tyszczuk, eds, *Culture and Climate Change: Recordings*. Cambridge: Shed, 75–93.

Hobba, Julian. 2018. 'Foreword.' In David Finnigan, ed., *Kill Climate Deniers*. Ebook. https://davidfinig.com/shop.

Lease, Bryce. 2020. 'Interview with Katie Mitchell.' *Contemporary Theatre Review* 30 (2), 253–59.

Love, Catherine. 2020. 'From Facts to Feelings: The Development of Katie Mitchell's Ecodramaturgy.' *Contemporary Theatre Review* 30 (2), 226–35.

Macmillan, Duncan and Chris Rapley. 2016. 'Note from the Authors.' In Duncan Macmillan, ed., *Plays: One*. London: Oberon.

May, Theresa J. 2021. *Earth Matters on Stage: Ecology and Environment in American Theater*. Abingdon: Routledge.

Merritt, Stephanie. 2014. 'Climate Change Play 2071 Aims to Make Data Dramatic.' *The Guardian*, 5 November. https://www.theguardian.com/stage/2014/nov/05/climate-change-theatre-2071-katie-mitchell-duncan-macmillan, accessed 1 May 2025.

Ogden, Emily. 2022. *On Not Knowing: How to Love and Other Essays*. Chicago, IL: University of Chicago Press.

Read, Alan. 2008. *Theatre, Intimacy & Engagement: The Last Human Venue*. Abingdon: Routledge.

Woynarski, Lisa. 2020. *Ecodramaturgies: Theatre, Performance and Climate Change*. Cham: Palgrave Macmillan.

23

JUST AND SUSTAINABLE FUTURES

How Will We Do It?

Anitra Nelson

Much media on climate change, the climate emergency and climate justice—including fiction and documentary films—is distressing. The top three films in Earth.org's top ten climate change movies for 2024 are illustrative (Earth.org 2024). Leonardo DiCaprio stars in *Don't Look Up* (McKay 2021), in which a metaphoric plot has scientists scrambling to draw attention and an adequate response to a comet on a disaster path targeting planet Earth (Earth.org 2024). *Princess Mononoke* (Miyazaki 1997) is a 'hard-hitting political anime' that 'depicts the consequences of destroying nature and humanity's inability to live in harmony' (Earth.org 2024). Similarly, *The Day the Earth Stood Still* (Derrickson 2008) is a narrative of paralysis: 'humans are destroying the Earth's environment, and the aliens have come to save the planet and its species, with the exception of humankind …' and a warning of what the future holds for us (Earth.org 2024).

This chapter explores the roles of audio-visual and other creative techniques as key supports for social transformation constructively addressing climate change. Through the lens of grassroots activism, it considers the theatre that society becomes in times such as ours, with performative oligarchs and protesters such as Extinction Rebellion featuring in the daily news. Transformation, agency and creativity are critical—rather than received scripts and directed parts of a receding past. Impromptu self-made theatre invents the future. The creative, educational and political short documentary *Beyond Money: Yenomon* came out in mid-2022 (Nelson 2022a), drawing from the book *Beyond Money: A Postcapitalist Strategy* released a few months earlier (Nelson 2022b). It was designed for educational purposes, a performing arts response to the climate emergency. My filmmaker's position highlights agency, Earth (nature) as a context and

DOI: 10.4324/9781003584445-30

source of life, bricolage and assemblage. To contain the discussion, the setting is the degrowth movement, its values, visions, relations and techniques. Let us now step onto that stage.

Climate Creativity as Grassroots Agency

Quality factual documentaries on climate change are needed, but 'you are just a grain of sand' messages inevitably weigh down audiences. For instance, a polished 58-minute documentary, *Climate Changers*, directed by Johan Gabrielsson (2023) follows eminent Australian scientist Tim Flannery in a global search for leadership. Both Gabrielsson's documentary and Earth.org's 2024 top three films cited above emphasise mainstream vertical power structures and a planetary scope with unintended effects of overwhelming viewers, setting up feelings of frustration and impotence. Even the call to action is couched in hierarchical capitalist organising terms: 'What can you do to be a climate leader?' (Climate Changers 2024).

In contrast, grassroots performing arts responses to the climate emergency pose stronger critical challenges to the capitalist economy and polity, along with greater senses of empathy and empowerment at the levels of both the individual and community. *Beyond Money* (Nelson 2022b) argues that we humans face two pressing challenges: various forms of ecological unsustainability and socio-economic inequities. Even if most discussion and action centres on global heating and containing climate emissions, carbon emissions are only the tip of the iceberg of the ecological crises. The term 'global heating' expresses Earth's potentially fatal fever. We can't just treat the fever; we have to address its cause. The orientation is towards action and hope. This is the 'dignified revolutionary rage' of Holloway (2022, back cover), who situates hope 'in our richness, in the joy of our collective creativity'.

Historically, anti-capitalist critics pointed to capitalism's social deficiencies, such as alienation, inequities and exclusion. But today, deep systemic damage to ecosystems is encapsulated in calls for 'system change not climate change' (Klimaforum09 2009), and it is widely acknowledged that global heating results from human, particularly capitalist, activities (United Nations 2025). What might system change demand? Most postcapitalists only question money to the extent of proposing a different type of currency, or preferencing distinctive ways of sourcing and using it, but a stream of non-market socialists explore non-monetary futures (Nelson 2022b). They include 'real valuist' arguments for a community mode of production—a world based on social and ecological values—with 'green materialists' such as degrowth activists, as agents of transformation (Nelson 2015). This real valuist approach underscores the messages and style of the short documentary in question.

To highlight its content and practical messages, in sympathy with the degrowth world it portrays, *Beyond Money: Yenomon* (Nelson 2022a) is simply narrated

FIGURE 23.1 *Beyond Money: Yenomon* flyer. Photograph by Bronwyn Silver. Image courtesy of Anitra Nelson.

behind a series of stills of landscapes and communities practicing self-organising and collective sufficiency. The documentary was created in 2022 by women living in Castlemaine (Central Victoria, Australia) and its surrounds. A low-budget film, it was directed, produced, written and narrated by the author of this chapter, who also contributed some of the stills. The documentary's achievements owe much to the spectacular macro photography of Bronwyn Silver, the highly skilful editing by Leonie van Eyk and a didgeridoo soundtrack played by its composer, the late Bruce Rogers (see Figure 23.1).

Degrowth: A Milieu

The degrowth movement arose most visibly in the last two decades as a philosophical and political challenge to economic growth, which has become a driving force across the world (Schmelzer et al. 2022). The movement questions not only why we work, produce and consume more than we need but also why we don't have an economy oriented around the simple satisfaction of needs. Such questions intrigued bohemian artists such as William Morris who railed against 'the war of commerce' and envisaged liberating futures full of creativity. Such precursors of degrowth strove for lives consumed by experiential, sensual and tactile being: 'Senseless luxury, which Morris knew cannot exist without slavery of some kind, would be replaced by communal luxury, or equality in abundance', writes Ross (2015, 63). Increasingly breaching the ecological limits of Earth and human limits of exploitation and degradation, degrowth has come of age as a concept and movement appearing in mainstream media. Still, degrowth is all too often woefully misconstrued either as a call for austerity and depression or as simply a movement concerned with minimising production and consumption, as modifying societal material throughput in terms of social metabolism.

In reality, degrowth is intellectually creative, turning growth on its head. Its advocates refer to a dramatic battle of 'imaginaries'—the imaginary of growth versus its decolonisation in degrowth. As such, degrowth is *qualitatively* distinct from growth. *Beyond Money: Yenomon* (Nelson 2022a) seeks to convey, and is imbued with, degrowth values and approaches. The degrowth movement highlights 'convivial technology'—tools such as a bike or a spade are simple, durable and appropriate, easy to access, understand, use and repair (Schmelzer et al. 2022, 156–57, 228–31). Convivial technology is in sharp contrast to images in Charlie Chaplin's *Modern Times* (1936) where workers churn out cheap commodities. In a more convivial and caring approach, *Beyond Money* is made in the artistic style of digital storytelling (ACMI 2017).

Degrowth activists find it essential to act out, practice and prefigure visions of an equitable and sustainable future. They challenge advertising in the style of 'adbusters', the global collective creating anti-consumer advertisements (see Figure 23.2). They take part in theatrical Critical Mass actions where bike

FIGURE 23.2 Spoof ad, created by Marc de Jong. Image courtesy of Adbusters Media Foundation.

riders disrupt traffic and create spaces free of mechanical vehicles prefiguring a future where people and bikes reign. Similarly, degrowth activists develop collective forms of designing, building and inhabiting housing, and collective ways of growing, caring for, harvesting and sharing food (Nelson and Schneider 2018; Nelson and Edwards 2021). To be a degrowth activist is to treat this world as a stage on which to construct the world anew in ways that attract attention, engagement, wonder and jog the imagination enabling people to enact new futures. The *Beyond Money: Yenomon* documentary was designed to flesh out how such futures might look in practice.

Prefigurative Hybrids

Degrowth activists exist in our current world but engage in prefiguring postcapitalist futures. The documentary begins with scenes of burnt landscapes and notes of money pegged to a clothesline, as the narrator talks of crippling debts and natural disasters. Capitalism is defined as money making more money, a carrot and stick, a basic meme for life today. Its scripts and stage directions revolve around working and buying and working to buy again. The wealthy rise and fall on the basis of monetary worth. '[I]f money go before, all ways do lie open', suggests Ford (as Brook) in Shakespeare's *The Merry Wives of Windsor* (Act 2 Scene 2, lines 169–70). Today, money organises us so we cannot self-organise. In contrast, degrowth promotes self-organisation on the basis of community, autonomy and subsidiarity.

This short documentary has been received as a version of how degrowth might evolve, even though its no-money angle is confidently pursued only by a minority within degrowth thinkers and activists (Nelson 2024a). Carbon emissions are directly related to overconsuming (growing extractivism and trade). We cannot cut production because 'the economy' might die. The stark fact is that capitalism, the market, has no operating systems to simply produce less. The processes that capitalists follow in producing commodities for trade mean that they have to create a buffer, profits, and produce as much as they can to ensure they keep solvent in a competitive market. The processes driving production for trade rely on constant growth, on competitive production for trade, which is always about winners and losers. Similarly, income earners must make sure they don't overspend and go into debt. Monetary dynamics intrinsically cause inequities. Inevitably, it's about expanding production, generating and regenerating inequity.

Given that the economy as we know it, a monetary economy, has no capacity to deliver social and ecological values, it is a necessary even if not sufficient step to move beyond it. While many people recognise the problems with money, one quickly finds that they often find it hard to imagine a money-free future. What might a satisfactory money-free economy look and feel like? It must address the two great challenges facing humanity. It must enable us to fulfil everyone's basic needs, neither more nor less. It must enable us to live within Earth's limits and protect it from the ravages and capitalism.

Practices embodying *prefigurative* politics aim to demonstrate imaginaries of preferred futures in the here and now (Monticelli 2021). Creating and circulating the low-budget *Beyond Money: Yenomon* represented a 'prefigurative hybrid' activity (Leahy 2021, 134–69). Prefigurative *hybrids* are, by circumstance, constrained by the capitalist contexts within which they exist as marks of resistance, simultaneously pointing towards equitable and sustainable ways of living in a degrowth future. This short documentary was produced using as many non-capitalist approaches and non-commodified forms as possible. Given the often market-oriented expectations to which grant-takers are obliged to conform, I didn't apply for funding. I, as the filmmaker, performed all roles—as script writer, producer, director, narrator and one of the photographers—without any monetary benefit. Indeed, I bore the monetary costs of remunerating the main photographer, the film editor and composer-performer, and costs associated with promotion and screenings, including entries to film festivals and certain other events. It has earned very little income because every effort was made to distribute it widely and for anyone to access a viewing freely. Moreover, although the book on which the documentary draws is commercially available in several formats, it is also accessible as a free PDF through the scholarly open-access funder Knowledge Unlatched.

The artistic means of making the short documentary needed to mimic its ends, so I chose to select a series of stills from photographs taken by me in the past of several communities that practice self-organising, that care about the planet and people and that practice permaculture and/or agroecology. Ordinary shots of people co-working in orchards are heightened in their effect by interlacing them with Bronwyn Silver's striking landscapes and ecologically close-up macro photography. One negative about such bricolage contributed to the only strong criticism of the film received. An international jury found it strange, indeed exclusive, that no people of colour were shown, even though the film included shots of multicultural Australians.

I worked alongside a remarkable local film editor Leonie van Eyk, who not only created an animation with scrabble-style letters to highlight some of the more abstract ideas in the documentary but was also responsible for recording the sound and determining the film's movement and pace. The didgeridoo soundtrack came from a CD of the late Bruce Rodgers, by permission of his partner. I used to live with both at the Round the Bend Conservation Cooperative, an intentional community in Kangaroo Ground/Christmas Hills on the peri-urban fringe of Melbourne. As such, the documentary was made through very personal connections and community-oriented collaborations.

Learning: Imagining, Discussing and Doing

In contrast to spectacular and overpowering blockbuster climate disaster documentaries, grassroots performing arts responses to the climate emergency are generally consciously designed towards community-based empathy and empowerment. The creative alliance of arts, humanities and sciences climate action advocates CLIMARTE, for instance, supports 'groundbreaking participatory exhibitions and creative projects addressing interconnected and interdependent ecological impacts as well as effective, justice based solutions' and offers 'ongoing opportunities for the community to collectively express deep fear, grief, anxiety and despair while organising actively hopeful pathways to urgently usher in available solutions' (CLIMARTE 2023, 7). Indeed, the short documentary in question was not designed for massive cinematic audiences but rather as a discussion-starter for smaller groups of participant-learners.

The aim was to provide a sense of community among viewers. The money-free 'world' portrayed is always introduced as illustrative rather than prescriptive. Viewers are invited to imagine their own distinctive money-free world. Being short and brief, the documentary gives participants a common reference point for discussion. Moreover, the viewer-cum-discussant has an opportunity to identify as an agent, is directed to acknowledging Earth as their source of life, and to consider futures replete with direct democracy. The milieu is an ecological locality and its abundance for fulfilling collective sufficiency using creative and constructive methods of bricolage and assembly to co-make futures that are sustainable and equitable.

The documentary shows how contemporary sustainable anti-capitalist alternatives such as community food gardens, maker labs and social centres can grow, multiply and interconnect as a micro-ecology of human activities within a macro-ecology of more than human life. The removal of money and market indicates transformation as metamorphosis. Commoning and caring are central to relationships that are no longer moulded and contorted by transactional dynamics (Smith 2024). At the risk of depriving viewers' imaginations from taking flight, the documentary offers images of the future to convey the auteur's arguments and messages. In these ways, the approach leans towards a holistic and ecological conceptualisation of utopia as a 'method' and 'speculative sociology of the future' (Levitas 2023, 468).

The imaginaries of non-monetary economies in *Beyond Money: Yenomon* (Nelson 2022a) have drawn on a number of years of lived experience of communities where activities that I advocate already happen (in prefigurative hybrid forms). In classrooms and public discussions (Nelson 2025), I've found this the easiest way to get discussion around futures of hope. The documentary is a call for everyone to put their shoulder to the wheel and centres the viewer's attention on key questions useful in mobilising change in everyday lives. Why do we need to transform? How will we do it? Could we—and would we really need to—live like that? If we can't, how might we achieve degrowth goals in a different way? The conversation immediately dives away from the ugly and seemingly intransigent truths of ecological collapse, pleading in humiliating ways with those in power, and of facing weak action by politicians, policymaker apathy and a nightmare barbarism. As if composting such depressive and anxiety-producing discourses, the messages in the documentary are designed instead to propel people to focus on 'doing', on the benefits of being active agents convivially creating a healing and regenerative future in which they are active agents.

Many imagine that the communities portrayed in the documentary are rare, cultish and small-scale while climate change is a global challenge. As both the macro photography and the book on which the documentary (Nelson 2022a) is made indicate, communities are tiny, permeable and replicating cells *scaling out* in ecological bioregional ways across Earth. Thus, the Yenomon proposal is glocal—universal and local (Nelson 2022b). Thousands of communities resisting capitalism based on traditional forms of co-governance and collective provisioning are further living examples spoken of in prefiguring terms by others such as Barkin (2025).

While winning educational awards and screening in AAG Shorts 2024, a special session of the American Association of Geographers (AAG) annual conference in Hawai'i (Film Geographers 2024)—the documentary was just as determinedly designed to show the *arts* of self-organisation, frugal abundance and convivial technology approaches at the heart of degrowth and climate action. Art awards at film festivals include those in categories of mood, social, no-dialogue, monologue and liberation/social justice/protest (Nelson 2024b). Textural, futuristic and realistic, low budget and highly aesthetic, it unfolds a holistic alternative way of being, acting and practising our lives—intending to position the viewer as an agent and our relations as impromptu acts in the veritable theatre of life.

References

ACMI. 2017. 'Collections Digital Storytelling.' *ACMI,* 11 December. https://www.acmi.net.au/stories-and-ideas/Collections-digital-storytelling, ACMI, 11 May, accessed 1 May 2025.

Barkin, David. 2025 'Latin American Indigenous Perspectives Meet Degrowth.' In Anitra Nelson, ed., *Routledge Handbook of Degrowth*. Abingdon, Routledge, pp. 153–165.

Chaplin, Charles, dir. 1936. *Modern Times*. [Film.] United Artists. Hollywood, Los Angeles (California).

CLIMARTE. 2023. *Strategic Plan: 2024–2026*. CLIMARTE. December. https://climarte.org/wp-content/uploads/2024/01/CLIMARTE-Strategic-Plan-2024-2026.pdf, accessed 1 May 2025.

Climate Changers. 2024. 'Take Action.' *Climate Changers*. https://climatechangersmovie.com/take-action, accessed 13 September 2025.

Derrickson, Scott, dir. 2008. *The Day the Earth Stood Still*. [Film.] 20th Century Fox. Vancouver Film Studios.

Earth.org. 2024. '10 Climate Change Movies to Watch This Year.' *Earth.Org*, 5 April. https://earth.org/climate-change-movies, accessed 1 May 2025.

Film Geographers. 2024. 'AAG Shorts 2024.' *Film Geographers*. https://www.filmgeographies.com/events/aag-shorts-2024, accessed 1 May 2025.

Gabrielsson, Johan, dir. 2023. *Climate Changers*. Totem Films/Antidote Films.

Holloway, John. 2022. *Hope in Hopeless Times*. London: Pluto Press.

Klimaforum09. 2009. *System Change—Not Climate Change: A People's Declaration from Klimaforum09*. Copenhagen: DGI-BYEN.

Leahy, Terry. 2021. *The Politics of Permaculture*. London: Pluto Press.

Levitas, Ruth. 2023. 'There's No Such Thing as "the Economy", Stupid: Using Utopia to Imagine Society "after Money".' *Review of Evolutionary Political Economy* 4, 467–79. https://doi.org/10.1007/s43253-023-00096-9.

McKay, Adam dir. 2021. *Don't Look Up*. [Film.] Netflix. Massachusetts.

Miyazaki, Hayao dir. 1997. *Princess Mononoke*. [Animation.] Toho. Studio Ghibli.

Monticelli, Lara. 2021. 'On the Necessity of Prefigurative Politics.' *Thesis* 11, 167 (1), 99–118.

Nelson, Anitra. 2015. 'New and Green Materialism.' *Progress in Political Economy*, 29 July. https://www.ppesydney.net/new-and-green-materialism, accessed 1 May 2025.

Nelson Anitra, dir. 2022a. *Beyond Money: Yenomon*. Anitra Nelson/Vimeo. https://vimeo.com/722765718.

Nelson, Anitra. 2022b. *Beyond Money: A Postcapitalist Strategy*. London: Pluto Press.

Nelson, Anitra. 2024a. 'Degrowth: Monetary and Non-Monetary Economies.' In Laura Eastwood and Kai Heron, eds, *De Gruyter Handbook of Degrowth*. Berlin/Boston, MA: De Gruyter, 41–53.

Nelson Anitra. 2024b. 'Creative Artist.' *Anitra Nelson*. https://anitranelson.info/creative-artist, accessed 1 May 2025.

Nelson, Anitra. 2025. 'Events.' *Anitra Nelson*. https://anitranelson.info/events, accessed 1 May 2025.

Nelson, Anitra and Ferne Edwards, eds. 2021. *Food for Degrowth: Perspectives and Practices*. Abingdon: Routledge.

Nelson, Anitra and François Schneider, eds. 2018. *Housing for Degrowth: Principles, Models, Challenges and Opportunities*. Abingdon: Routledge.

Ross, Kristin. 2015. *Communal Luxury: The Political Imaginary of the Paris Commune*. London/Brooklyn, NY: Verso.

Schmelzer, Matthias, Andrea Vetter, and Aaron Vansintjan. 2022. *The Future Is Degrowth: A Guide to a World beyond Capitalism*. London: Verso.

Smith, E.T. 2024. 'Practising Commoning.' *The Commons Social Change Library*. https://commonslibrary.org/practising-commoning/, accessed 1 May 2025.

United Nations. 2025. 'What is Climate Change?' *United Nations: Climate Action*. https://www.un.org/en/climatechange/what-is-climate-change, accessed 13 September 2025.

INDEX

Note: *Italic* page numbers refer to figures and page numbers followed by "n" denote endnotes.